Multi-Microprocessor Systems for Real-Time Applications

International Series on
MICROPROCESSOR-BASED SYSTEMS ENGINEERING

Multi-Microprocessor Systems for Real-Time Applications

edited by

GIANNI CONTE

and

DANTE DEL CORSO

Department of Electronics,
Polytechnic of Turin, Italy

D. REIDEL PUBLISHING COMPANY

A MEMBER OF THE KLUWER  ACADEMIC PUBLISHERS GROUP

DORDRECHT / BOSTON / LANCASTER

Library of Congress Cataloging in Publication Data

Main entry under title:

Multi-microprocessor systems for real-time applications.

    (International series on microprocessor-based systems engineering)
    Includes bibliographies.
       1.   Real-time data processing.   2.   Multiprocessors.   3.   Computer
architecture.   I.   Conte, Gianni, 1946–   II.   Del Corso, Dante,
1946–   III.   Series.
QA76.54.M85   1985   001.64′4   85–10695
ISBN 90–277–2054–1

Published by D. Reidel Publishing Company,
P.O. Box 17, 3300 AA Dordrecht, Holland

Sold and distributed in the U.S.A. and Canada
by Kluwer Academic Publishers,
190 Old Derby Street, Hingham, MA 02043, U.S.A.

In all other countries, sold and distributed
by Kluwer Academic Publishers Group,
P.O. Box 322, 3300 AH Dordrecht, Holland

CONTENTS

CONTENTS

PREFACE

The continous development of computer technology supported by the VLSI revolution stimulated the research in the field of multiprocessors systems. The main motivation for the migration of design efforts from conventional architectures towards multiprocessor ones is the possibility to obtain a significant processing power together with the improvement of price/performance, reliability and flexibility figures. Currently, such systems are moving from research laboratories to real field applications. Future technological advances and new generations of components are likely to further enhance this trend.

This book is intended to provide basic concepts and design methodologies for engineers and researchers involved in the development of multiprocessor systems and/or of applications based on multiprocessor architectures. In addition the book can be a source of material for computer architecture courses at graduate level. A preliminary knowledge of computer architecture and logical design has been assumed in writing this book.

Not all the problems related with the development of multiprocessor systems are addressed in this book. The covered range spans from the electrical and logical design problems, to architectural issues, to design methodologis for system software. Subjects such as software development in a multiprocessor environment or loosely coupled multiprocessor systems are out of the scope of the book.

Since the basic elements, processors and memories, are now available as standard integrated circuits, the key design problem is how to put them together in an efficient and reliable way. Therefore, the book puts a special enphasis on the interconnection structures, covering both design and analysis aspects.

Chapter 1 (Multiprocessors System Architecture, by P.Civera, G.Conte, and F.Gregoretti) contains a general introduction in the subject of multiprocessing and describes some existing architectures. Chapter 2 (Performance Evaluation of Multiprocessor Systems, by M.Ajmone Marsan, G.Balbo, and G.Conte) provides some tools for the evaluation of the performance of bus oriented multiprocessor systems. Chapter 3 (Software Design for Multiprocessor Systems, by F.Gregoretti) analyzes the problems related with the implementation of kernel firmware in multiprocessor systems. Chapter 4 (Design of Multiprocessor Buses, by D.Del Corso) gives a methodology to analyze and design parallel buses, while Chapter 5 (Some Examples of Multiprocessor Buses, by P.Civera, D.Del Corso, and F.Maddaleno) describes some existing standards. Chapter 6 (Hardware Modules for Multiprocessor Systems, by D.Del Corso, and M. Zamboni) discusses the design of the basic building blocks for multiprocessor systems. Chapter 7 (Multiprocessor Benchmarks, by E.Pasero) presents some experiences of parallel processing on a multiprocessor machine.

This book is one of the results of a 5-years research effort on multiprocessor architectures carried on by the authors in the frame of the MUMICRO project of the Italian National Research Council (C.N.R.) Computer Science Program. The authors are members of the Dipartimento di Elettronica of the Politecnico di Torino, except G.Balbo, who belongs to the Dipartimento di Informatica of the Università di Torino.

We would like to thank Prof. Angelo Raffaele Meo and Prof. Ugo Montanari, heads of the above mentioned Computer Science Program, for the support given to the project. We are expecially indebted to Prof. Roberto Laschi, coordinator of the MUMICRO project, for his valuable advice and encouragement during the whole duration of the work. We vould also thank the friends and colleagues of both MUMICRO and MODIAC projects for their comments and suggestions.

The Dipartimento di Elettronica of the Politecnico di Torino provided in these years the background that made possible the development of the research activity and the support for the preparation of the book. The editing wase made easier by the help of Patrizia Vrenna (typing) and Luciano Brino (drawings). The suggestions of Ian Priestnall of Reidel Co. helped the final preparation of the book.

Gianni Conte - Dante Del Corso

Torino, March 1985

CHAPTER 1

MULTIPROCESSOR SYSTEM ARCHITECTURE

P.Civera, G.Conte, F.Gregoretti
Dipartimento di Elettronica
Politecnico di Torino
Torino ITALY

ABSTRACT.   In this chapter the different architectures belonging to the class of distributed systems are first presented. Multiprocessor architectures are then focused upon, because they represent the area of main interest of the book. The graphic notation that will be used in this chapter and in the following ones to describe the different multiprocessor architectures is presented. An overview of the more significant multiprocessor systems is then included; the TOMP multiprocessor, developed by the authors, in cooperation with others, is presented in greater detail.

1.1. DISTRIBUTED PROCESSING AND MULTIPROCESSORS

1.1.1. Classification Criteria

Multiprocessor systems are part of the large class of "Distributed Computing Systems". Many classifications and taxonomies have been published on this subject, but some different views still exist together with disagreements on what can be considered a "Distributed Computing System". This section is not an extensive review of the field but briefly presents the most commonly used classification criteria and the related terminology. This will help one both to get a general view of the field and to classify in a correct frame the systems whose design criteria are presented in the following sections.
      A first way to classify the different types of distributed systems can take into account the "granularity" |JONE80| of the interaction among the activities that are executed in parallel by the system. It is possible to consider systems on which the cooperation (data exchange and/or synchronization) among the

1

G. Conte and D. Del Corso (eds.), Multi-Microprocessor Systems for Real-Time Applications, 1–31.

elements occurs very seldom and involves large blocks of structured data at one time or, on the other hand, distributed systems on which the data exhange is very frequent and the synchronization occurs, for example, at the instruction level.

1.1.2. Computer Networks

The Computer Network is the oldest class among distributed systems and took its origin from the connection of large mainframes. Each processor retains a strong local autonomy and dedicates only a limited part of the processing power to common activities. For this reason there is no general agreement about considering computer networks as a part of distributed processing. The rapid spreadout of mini/microcomputer systems and the availability, at low cost, of super-minis has created a new interest in this field, by the introduction of "local networks" of many small computers. This has changed in some aspects the philosophy of the computer network. The main characteristic of a network is still maintained, being the fact that each processor is a stand alone computer and their processing activities are independent. On the other hand, global policies for the handling of common resources, like high cost peripherals or distributed data bases, are introduced and processors "cooperate" in their management.

1.1.3. Multiple Processor Systems

The second area refers to the Multiple Processor systems in which each processor is a fully programmable unit that can execute its own program, but the set of processors forms a single entity. Therefore the issue which distinguishes computer networks from multiple processor systems is the fact that in this latter case all the system resources are coordinated towards a common task with a single (centralized or distributed) control mechanism. The amount of information exchange among the basic processing units can now be significantly greater than in the previous case. The interconnection topology and the communication strategy among the processing elements become in this case the crucial point of the system and a more detailed classification must be based on the structure of the interconnection network.

The communication network connecting the processing elements becomes more general and complex and a further broad classification is based on the ability of these interconnections of supporting (or not) the sharing of an address space between processors. Systems in which the processing elements do not share memory and are connected through I/O data links are in general defined as MULTIPLE COMPUTER or LOOSELY COUPLED systems whereas structures with a common address space are called MULTIPROCESSORS or TIGHTLY COUPLED systems. The architectures of the first class may be, depending on the size, similar to that of a geographically distributed computer network. The interconnection network can be made using a parallel or a serial link, and the data transmission rate can range from few kbit per second up to 10 Mbit per second. The second class comprises all the systems on which a number of processors can access a common memory area.

1.1.4. Special Purpose Machines

The third area refers to special purpose machines, that is processing systems designed to solve a given problem or fields of applications. These structures can be stand alone machines or can be connected, as peripherals, to some high power computing systems such as mainframes, in order to speed-up some frequently needed specific operation. In this class we can find:

1. High Parallel Structures.
   They are composed of a large number of identical hardware units, each one able to perform a fixed basic operation. These units are connected together and work in parallel for the fast solution of computer-bound algorithms like matrix operations, or discrete Fourier transforms, provided that convenient algorithms could be found. An example of high parallel computing structures are the systolic array whose architectural properties seem very suitable for VLSI implementation |MEAD80|. A systolic system consists of a set of identical computing cells interconnected according to a regular topology in which the flow of information is allowed only among adjacent units in a pipeline style. The I/O needs for the single unit are so limited and the shortness of the interconnection allows a significant speed-up of the operations. It is therefore satisfied one of the major constraints of VLSI elements on which the computing capabilities depend on the number of the active elements, and therefore to the area of the silicon, whereas the number of interconnections is limited by the length of the border.

The programmability of these structures is extremely low because
each one is specifically designed and optimized to carry out
efficiently only a well defined algorithm.

2. Array Processors
   Array Processors perform in lockstep the same operation on
   several different data. These machines, defined also as Single
   Instruction Multiple Data (SIMD) structures, have a higher
   degree of programmability than the previous ones, but their use
   is restricted to problems with a high and finely grained
   parallelism such as the manipulation of large array of data
   types.

3. Non Von-Neumann Machines.
   Von-Neumann machines are characterized by the presence of a
   processing unit that executes instructions (stored in memory) in
   sequence under the control of a program counter. The sequential
   program execution does not allow an efficient exploitation of the
   parallelism that can be inherent to the program. One of the
   proposed architectures suggested to overcome this problem is the
   data flow computer; in this case the execution of an instruction
   is allowed as soon as the requested operands (and the hardware
   resources) become available. This type of architecture is also
   called data-driven, whereas Von-Neumann machines are indicated
   as control-driven systems. It must however be pointed out that
   usually Non Von-Neumann machines are obtained by connecting
   together, in some peculiar way, elements that operate in a Von-
   Neumann style. This is the reason why they are here considered
   in the large family of distributed system.

1.1.5. Other Classifications of Distributed Systems

The above classification is based on the granularity of the
interaction among the units composing the systems. Different
classifications can be found in the literature. Among these the more
significant ones were proposed by Flynn |FLYN72|, and by Enslow
|ENSL78|. Flynn introduced the following three classes of computer
organization:

1. The "single-instruction stream, single-data stream" (SISD) which
   represents the conventional uniprocessor computer system.

2. The "single-instruction stream, multiple-data stream" (SIMD) which includes array processors.

3. The "multiple-instruction stream, multiple-data stream" (MIMD) which includes most of the multiprocessor systems.

Enslow proposed the use of a three dimensional space to characterize the distributed systems:

1. The distribution of the processing units: it corresponds to the physical organization of the hardware structure that can go from a single central processor unit up to a geographically distributed multiple computer system.

2. The organization of the control: it can span from a system with a fixed control origin to a distributed system composed of a set of fully cooperating and homogeneous processing units.

3. The distribution of the data: it is possible to have systems with a centralized data structure and systems with a complete partitioned data base.

It can be difficult, in all real cases, to classify a system using one of the previously mentioned schemes. For instance, in a multiprocessor system the level of the cooperation among the activities is not only defined by the architecture of the system, but also by the operating system, possibly distributed, running on it or by the application program itself. The processors can, for example, run tasks executing independently and very seldom needing data exchanges, using a message passing scheme, like in a LAN structure, even if they have access to shared memory areas. The same system can, on the other hand, support pipeline operation on a single data stream with a very frequent exchange of intermediate results, like in a SIMD structure operating synchronously on vectors or arrays.

    In conclusion the aim of these classifications is not to offer a precise scheme on which to be able to insert a known architecture, but to offer a global view of the design space and of the possible different solutions in the frame of the large area of distributed computing systems.

1.2. MULTIPROCESSOR SYSTEMS.

1.2.1. Multiprocessor Structures

In this section we will explore the multiprocessor system area in order to point out the main advantages of this class of structures and to try a further classification |BOWE80, PARK83, HWAN84|.
        The more general structure of a multiprocessor system is depicted in Figure 1.1. A multiprocessor system consists of a set of master modules (such as processors), and of a set of slave units (memories and/or I/O modules) connected together by means of an interconnection structure. More generally the master units are the elements in the system allowed to issue an access request to the interconnection structure in order to perform data transfers; the slave units receive the request access from the master units and can accept and honour them. It must be pointed out that the actual direction of the information transfer can either be the same as the access request (write operation) or its opposite (read operation).

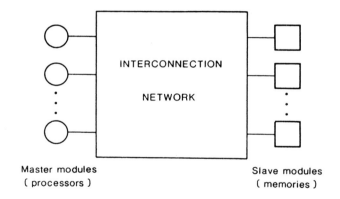

Master modules   Slave modules
(processors)   (memories)

Fig.1.1 – The most general structure of a multiprocessor system.

Master units other than processors can be, for example, I/O modules with direct memory access (DMA) facilities, but in the following we will very often use the term processor instead of master unit, as well as memory instead of slave unit. In this figure, and in the following ones, master units (or processors) are represented by circles and slave units (or memories) by squares.
The interconnection structure is the most important part of the system because the data exchange among the processing units depends on it.

No generally accepted standard solution already exists for the interconnection structure, and for this reason, whereas processors and memories are available at low cost as integrated circuits, the interconnection network is always designed according to the needs of the specific application or system.

1.2.2. The Interconnection Network

The minimal functional unit that can execute a program consists of a processor and a memory. The objective of the interconnection network is that of coupling at a given instant of time each processor with the requested memory module. Two main reasons can hinder the satisfaction of the processor requests:

1. two or more processor requests for the same memory unit.

2. two or more processing units need the same communication link to access different memory units.

In both these cases the processor that cannot access the requested memory unit must wait. In the first case the waiting time can be eliminated only by using memory modules with peculiar features, where for instance multiple read or independent read/write operations can be allowed. In the second case the structure of the interconnection network can reduce the time lost by the processors waiting for non-free communication resources. The complete (or almost) complete elimination of any sort of contention corresponds to the set up of a very expensive interconnection structure. The usual design challenge is the reduction of the complexity of the interconnection network without affecting the performance of the multiprocessor system.

One of the first design choices, from the architectural point of view, is the selection between a completely homogeneous set of memory modules and a hierarchy of specialized memory elements. The second alternative is of course the more effective one but it reduces the regularity of the system and may imply (from the user or the programmer point of view) a too detailed knowledge of the system behaviour. One of the more effective solutions is the partition of the global set of memory modules into two main groups. The first one is a set of independent memory modules each one associated to a processor and only accessible from it (private memory), these memories can retain, for example, the programs each processor will execute.

The second one is a set of common memory modules containing
the information that each processor in the system can access (with
the same or different accessing rights). The general structure of a
multiprocessor system with this architecture is shown in Figure 1.2.
From the point of view of the interconnection network different
solutions exist; we will therefore examine in greater detail: crossbar switches, multiport memory systems, shared bus systems, and
multistage networks.

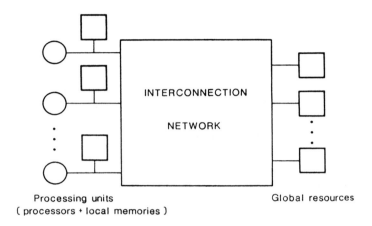

Processing units   Global resources
(processors + local memories)

Fig. 1.2 – A general structure of a multiprocess system.
Processing units are composed of a processor and of
a memory module accessible only from the processor.

1.2.3. Shared Bus

From the logical point of view this is the simplest interconnection
structure between one or many processors and memory modules. A
shared bus is a single communication path to which the functional
units (such as memories and processors) are connected, as shown in
Figure 1.3. If only one master unit is connected to the bus no
contention problems arises. When two or more master units are
connected to the same bus some policy must be used to establish the
link: a fixed time slot can be assigned to each master unit (or
processor), or the system may be able to resolve contention arising
from unconstrained request sequences. In the case of no fixed time
slice the processors require access to the memory modules through
an arbitration mechanism which handles simultaneous requests.

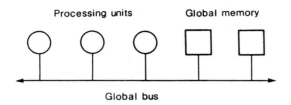

Fig. 1.3 – Single global bus multiprocessor system

It is obvious that this interconnection network does not allow simultaneous transfers between different processor/memory pairs and therefore the single bus structure may easily become the bottleneck of the whole system.   To get a better performance figure the single global bus can be substituted by a set of buses.   This approach is far more complex and a careful trade-off must be made between the cost,   the complexity,   and   the   resulting   functional   upgrading obtained with the redundancy.

1.2.4. Multiport Memory

In these systems the control and the arbitration logic,   that in a single global bus reside either on the processor or in the bus interface modules,   are concentrated in the memory modules that present a number of communication interfaces (ports) through which it is possible to access the internal information.
        From a logical point of view this approach gives no new architectural solution with respect to multiple shared bus systems, The interest toward this type of architectural solution will certainly grow up as soon as dual (or multi) port memory will be available as integrated circuits.

1.2.5. Crossbar Switches

In such a system a set of separate paths is connected to each memory bank and another one to each processor, as shown in Figure 1.4. A set of switches may connect any processor to a memory path. The system supports simultaneous accesses towards all memory units. Contention may arise only when the same memory   bank is requested by several processors at the same time.

The processing units may be composed by the CPU only or by the CPU with its associated private memory. The complexity of the interconnection network has limited, up to now, the use of this structure even if one of the earliest multiprocessor system, the C.mmp, has implemented this architecture using PDP11 minicomputer as basic processing units |WULF72|. The complexity and cost reduction of the basic switch is, in this case, a key point, therefore recent proposals suggest the use of VLSI elements to implement it |MCFA82|.

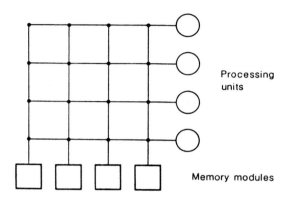

Fig. 1.4 – The basic structure of a crossbar multiprocessor system

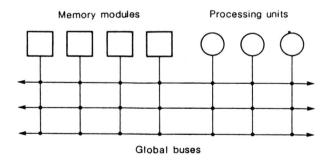

Fig. 1.5 – An example of multiple bus multiprocessor system

The generalization of the crossbar architecture is the multiple shared bus structure that is shown in Figure 1.5.
This network comprises two sets of elementary switches; the first one is the link between the processors and the bus, the second one

represent the links between the buses and the memory modules.   In
the general case the number of switches is (m+p)*b, where b is the
number of buses.   The crossbar architecture needs instead m*p
elementary switches.

1.2.6. Multistage Interconnection Networks

The interconnection network of the multiprocessor system can be
made using an array of modular building blocks of only one type
|FENG81|.   Each element can perform a very simple circuit switching
function.   Consider for example the basic 2 × 2 switching element
shown in Figure 1.6.

Fig. 1.6 – The two setting of the basic 2x2 switching element

The switching element can be set in two configurations performing a
direct and a crossed connection.   An array of N basic switching
elements (as shown in Figure 1.7) can perform a single stage
interconnection network.   A matrix of $N*log_2 N$ (base 2) basic
switching elements can interconnect a set of N input terminals to a
set of N output terminals.   Different interconnection strategies
between the stages generate different types of interconnection
networks.   In all these cases a convenient setting of the basic
switching elements can connect any input terminal to any output
terminal.
        In the case of multiprocessor systems the input terminal can
be a processing element (a processing unit with the associated local
memory) and the output terminal can be a global memory element.
If more than one terminal pair must be connected simultaneously
conflicts can occour in the communication path.   From this point of
view multistage interconnection networks can be divided into three
classes: blocking,   rearrangeable,   and non–blocking. Networks are
referred to as blocking if conflicts may occur.   A network is called
rearrangeable if,   by rearranging the existing connections,   it is
possible to establish a new interconnection path.

An   interconnection   network   on   which   all   possible
interconnections can be established without conflicts is called non-
blocking.

From a functional point of view these networks can also be
divided into those that allow information exchange on a circuit
switching or on a packet switching base. The first ones establish a
fixed connection between the input and the output port as long as
the data exchange occurs.   In a packet switching network a given
amount of information (a packet) is transmitted through the
network.   The packets are stored at intermediate points along the
path where they can wait if the path to the final address is not
free.

In the case of crossbar structure this interconnection network
will   be   really   effective   only   when   the   basic   element   becomes
available as VLSI chip.

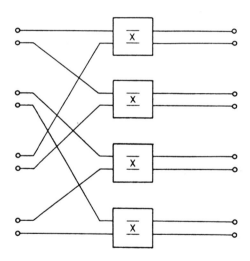

Fig. 1.7 – A single interconnection stage of a
shuffle exchange network

1.2.7. Applications of Multiple Processors

The potential advantages of multiple processor systems and the
motivations for their development have been expressed for a long
time by the following keywords: Good price/performance figure,
Flexibility, Extendability, High Availability.

It is now clear that multiprocessor systems will not offer an easy solution to all these problems, which can hardly be covered by single processor architectures. Many reasons can complicate the design and the use of a multiprocessor architecture. First of all the microprocessors now available as LSI or VLSI integrated circuits are not specifically designed to be used in a multiprocessor environment. Some new ones, now announced, have special features that will ease their use in a distributed environment. The second point is that there are no tools available for developing an application program and debugging it on a multiple processor system, or, if there are, they can be used only on some specific prototype systems. Moreover, nor the partitioning of complex problems into smaller activities that could be executed in parallel, nor the allocation of these activities on the physical structure of the distributed system are trivial tasks.

It must also be observed that fault–tolerance or high availability characteristics are not directly satisfied by a multiprocessor architecture but require special attention during all the phases of the design. Nevertheless the applications of multiple processor structures are increasing, together with the availability of systems whose internal architecture is essentially that of a multiprocessor. The main reasons are either the possibility offered by the multiprocessor systems of exploiting the physical distribution of the systems to be controlled, or the intrinsic parallelism of the class of algorithms to be executed.

In the first case we can mention all the systems devoted to plant or power control, to shipborne or airborne systems, to flexible work stations, to robotics, etc. To the second belong systems devoted to image and signal processing, to speach understanding, computer graphics, simulation, etc.

1.3. DESCRIPTION TECHNIQUES FOR MULTIPROCESSORS

1.3.1. Levels of Description

It is mandatory, before describing any complex system in detail, to clearly define the level of abstraction used for its description. According to a known structured top–down approach, a complex system may be partitioned into submodules and this procedure may be recursively repeated down to the most elementary blocks. A level of description is defined by the objects that are visible at that level and by the primitive operations that can be performed by or on them.

We can identify, starting from the more abstract ones: virtual levels, functional levels and physical levels.

VIRTUAL

At the virtual level the visible objects are the processes, the data structures on which they act and the language primitives used by the processes to manipulate data or to communicate with each other. This level is used in the description of virtual software environments, such as Concurrent Pascal machines or ADA run time supports. In this case the architecture of the underlying physical machine is in general not relevant. The evolution of technology allows the introduction at silicon level of operating system primitives and hardware support to the run time system. So this one may be, in the future, the only level at which the user will be allowed to interact with the system. An example of this trend is the iAPX 432.

FUNCTIONAL

A functional level allows one to see, at a very high level of aggregation, the properties of the module on which the physical system is divided. For example we can consider the memory as a unique module on which some operations such as read or write are allowed. At this level the relevant properties of every module are the set of functions they perform. In a processor system the visible objects at this level are the processors, the common resources and the logic interconnection topology.

PHYSICAL

Going into more detail a description of the system can be made specifying the logical and/or the physical implementation. Different approaches are in this case feasible, ranging from sophisticated Hardware Description Languages (HDL) at register transfer level down to the logical scheme describing the actual implementation.

1.3.2. Selection of the Description Level

From a general point of view, the purpose of a descriptive tool is to facilitate the communication of ideas in the frame of a defined area or subject. In the case of multiprocessor systems, description tools can be used and they are significant at the different levels of detail described in the previous section. At the very beginning of the design phase of a distributed system it is mandatory to use a descriptive tool in order to analyze the organization of analogous existing systems. The description obtained must be clear and unambigous, and must allow the comparison of the different solutions and the extraction from them of the most significant features. The same tool is, in consequence, the favourite candidate to describe the architecture of the system to be designed.

For this purpose the level of description must not go into details such as bus protocols, CPU instruction set and actual hardware implementation, but must offer an overall view of the system. According to the previous section, we shall call this level of detail FUNCTIONAL. A multiprocessor system is here viewed as a set of elements (such as CPU, memory, etc.) connected together. The importance of this level of description is relevant because:

- It allows a global view of the system; more details can be examined later, considering only one basic block at a time.
- It allows, in the first phase of the design specification, an easy information exchange among all the people involved in the work.
- It is the most convenient level of description for the performance evaluation when one consideres the analysis of the flow of data exchanged in the system.
- It is significant where, owing to the prevalent use of commercially available CPU's, the internal architecture of some of the more complex blocks is already defined.

1.3.3. The PMS Notation

The first and perhaps the most significant approach to the description of computer system at this top level was proposed by Bell and Newel |BELL71| and is known as PMS notation. The PMS description is basically a formalized graphical way to describe the architecture of a computer system. The term PMS, from Processor, Memory and Switch, allows one to understand easily the level of detail to which the description is dedicated. Other blocks can however be defined such as transducers, terminals, mass-memory, and so on.

The basic primitives for the PMS (defined by the functions they perform) are:

- Processor (P): a module capable of performing a sequence of operations that is executing a program.
- Memory (M): a module that can hold information.

– Switch (S):   a module making it possible to connect different
  modules in the computing system.
– Control (K): this module commands and/or supervises the operation
  performed by the other units.
– Transducers (T):   the module essentially changes the encoding of
  the informations and allows the I/O operation towards the
  external world.

PMS has evolved from the previous graphical form into a formal
language used to describe, in the frame of a large and coordinated
set of tools, the interconnection topology of computing structures
|BRAD79|.
      While PMS is devoted to the interconnection level, other
languages (i.e. ISPS |BARB81|) can be used to describe the
behavioural aspects of the same modules. In this case the ﹍two
languages cover complementary aspects (behavioural and structural)
of computing systems. It can be observed however that ISPS can
also be used to describe connections, but the use of PMS, in this
case, is far more efficient. Figure 1.8 shows the

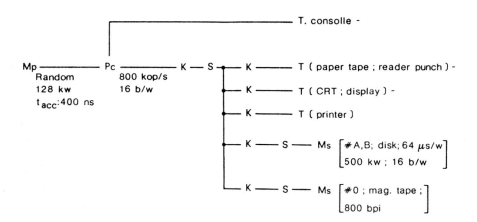

Fig. 1.8 – Description of a processing unit using the
PMS graphic language

description of a processing unit using the PMS graphical language,
while Figure 1.9 shows the description of a multiprocessor system.
Other details are added to each description to help the reader to
understand the function performed by the different modules.

The main drawbacks of the PMS description are:

- It describes only the interconnection topology of the system,   the
  behavioral aspect is left completely to other levels.
- The interconnection network does not specify the direction of the
  information flow and the origin of the control.

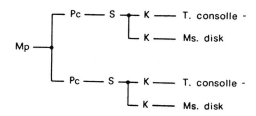

Fig. 1.9 – Description of a multiprocessor system using
the PMS language

1.3.4. The MSBI Notation

In order to overcome the main drawbacks of the PMS description we
introduced a new graphic model.   We derived from PMS the level of
the description and we added some new features mainly to improve
its effectiveness in the case of multiprocessor systems.   We consider
a multiple processor system as a set of modules exchanging "access
requests".   An access request fires a data transfer between two
modules.   For instance, a data transfer from memory to a CPU
register is triggered by an access request issued by the central
unit towards the memory specified by the address associated with
the request.   This request has a specific direction: from the module
starting the operation (in our case the CPU),   towards the modules
that receive and accept the transfer requests (memory).   It must be
pointed out that the actual direction of the information transfer can
either be the same as the access request (write operations) or
opposite (read operations).
In the following the word "direction" always concerns the
request. The basic building blocks used in the description are
shown in Figure 1.10.

MASTER: this module issues access requests;   the requests are
               sent within one or more specified address range(s).

SLAVE: this module receives access requests;   a slave accepts
                    and honours requests within one or more specified
                    address range(s).

BUS: this module supports the communications within the
                    other modules connected to it.

INTERFACE: this module transfers an access request between two
                    buses. It behaves on the one side like a slave, and on
                    the other like a master.   The interface has a
                    translation rule for the address from the slave to the
                    master side. As stated before, for what concerns the
                    access requests, the interfaces are unidirectional.

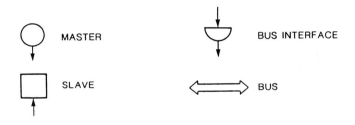

Fig. 1.10 – Basic elements for the MSBI descriptions

Since more than one master can be connected to the same bus,   an
arbitration mechanism is mandatory on each when more than one
master module is present.   Contention arising when masters
requesting access to a common resource need the same channel (bus)
can therefore be resolved.   The arbitration mechanism can be
distributed among the masters connected to the bus or implemented
as an "ad hoc" module. Nothing is said about the mechanism and/or
the policy of arbitration.   This information can be, if necessary,
informally added to the graphical representation.

1.4. SOME MULTIPROCESSOR SYSTEMS

1.4.1. Selection Criteria

In this section some multiprocessor system are described,   this
insight will not be exhaustive, but will try to show samples of all
kinds of implementation on this field.   The MSBI notation will be
used to describe the different architectures.

The selection is mainly made according to the different connection topologies:

- bus based multiprocessor (Cm*, μ*);
- cross-bar systems (C.mmp, Intel 432);
- a non-complete cross-bar system (PLURIBUS);
- a dual port memory based system (TOMP).

1.4.2. The Cm*

First of all we shall examine the Cm* multiprocessor developed at Carnegie Mellon University |SWAN76|. Cm* is composed of many clusters of processors connected together by an intercluster bus; each cluster contains several computer modules, connected via a Map bus; each computer module therefore is composed of one LSI 11 processor, 12 kword memory and related I/O interfaces. Figure 1.11 shows the description of the Cm* in PMS language.

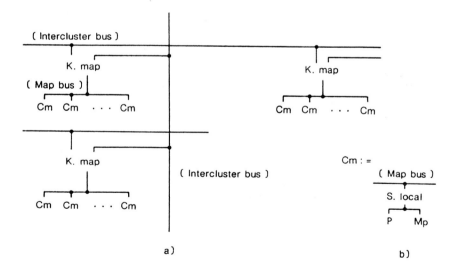

a)   b)

Fig. 1.11 – PMS description of the Cm* multiprocessor system
a) system structure
b) organization of a computer module

Up to 14 computer modules (Cm), each one composed of one processor (P) and some memories (M), are connected through a bus switch

(S.local) to the multiprocessor cluster bus (Map bus) forming a
Cluster.   Each   cluster   is   then   connected,   via   another
switch/controller (K.map), to 2 intercluster buses.

       Figure 1.12 gives the description of Cm* using the MSBI
notation; the dotted lines enclosing some modules are added only to
relate the description with the further one based on the building
block. Dotted lines also give an idea of the complexity of K.map
and S.local.

       The representation by means of MSBI clarifies the logical data
path through the several buses without compromising the simplicity
of the description. On the other hand this description loses any
relation with the physical building blocks of the computer machine;
in fact each block may contain more physical modules or conversely
only a part of it.

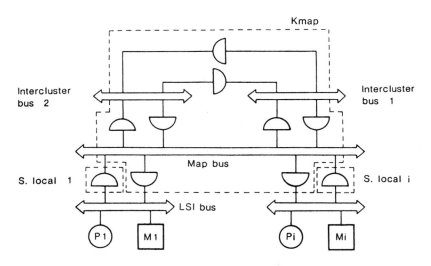

Fig. 1.12 – MSBI description of the Cm* multiprocessor system

1.4.3. The C.mmp

C.mmp is another multiprocessor system developed at CMU |WULF72|.
The C.mmp is a classic example of a cross-bar architecture.   In
Figure 1.13 it can be shown that, as in any crossbar architecture,
there are two kinds of buses: processor buses and memory buses.
These buses are arranged in rows and columns and the connections,
called Switch, are placed at each cross-point.

A single processor is connected to each row bus; these buses have only one processor placed on each of them, consequently they do not require any arbitration mechanism for access control.

The processor buses support local resources such as dedicated memories and peripherals. The column buses contain the shared resources of the system; they can be accessed by all the processors connected on the row buses, so they are multimaster and must be arbitrated.

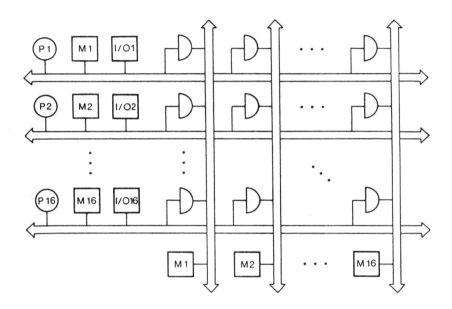

Fig. 1.13 – MSBI description of the C.mmp multiprocessor system

A C.mmp multiprocessor system was competely developed and was the object of heavy investigations on fault behaviour at 'system level', considering the fact that a multiprocessor system may also be viewed as a fault tolerant structure.

The developed C.mmp system was composed of 16 Digital Equipment PDP 11/40 CPU modules, 16 global memory modules of 32 kword capacity and a 16x16 switches crossbar matrix.

1.4.4. The PLURIBUS

Another well known solution for the interconnection scheme in a multiprocessor architecture is represented by the PLURIBUS system

|KATS78|. PLURIBUS is an operational multiprocessor system used as
interface message processor (IMP) on the ARPA network. Design
goals for the PLURIBUS were the maximum size-flexibility and the
highest reliability at the best cost-performance trade-off. In order
to obtain maximum size-flexibility all the processors are identical;
to obtain high reliability the whole system is redundant and
presents no common point of failure, moreover all the modules can
be physically isolated to protect the system against failure
propagation. In Figure 1.14 a MSBI representation of the PLURIBUS
architecture is given.

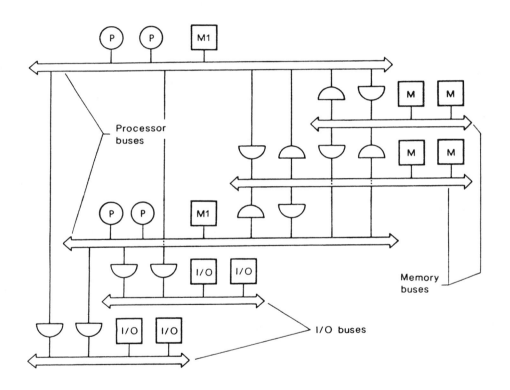

Fig. 1.14 – MSBI description of the PLURIBUS multiprocessor system

PLURIBUS is composed of three kinds of communication path:

    1) processor buses;
    2) memory buses;
    3) peripheral buses.

Each processor bus contains one or two processors. Each processor bus is connected to memory and I/O buses. The memory and the I/O buses contain shared memories and common I/O interfaces.

The architecture of PLURIBUS refers to a "distributed crossbar switch" system, where each switch element, called Bus Coupler (BC), is split into two boards and then placed on the two interconnected buses. Generally the "distributed crossbar matrix" in PLURIBUS is not complete. Bus Couplers act also as address mapping elements, the Bus Couplers map the 16 bit processor address space into a 20 bit system address space.

Each communication path is physically independent of the other ones. In order to obtain a graceful degradation of the system, under a single failure, or to allow on-line maintenance, at least two independent paths are always present. The PLURIBUS sytems use as processor element the SUE minicomputer, a 16 bit machine developed by Lockheed, and 32k to 80kword memory on each bus.

1.4.5. The μ^* System

The μ^* system is a multimicroprocessor developed at Politecnico di Torino, based on simple low cost microprocessors used as building blocks |CIVE82|.

The μ^* architecture is similar to the cluster part of the Cm*. Essentially μ^* is composed of several identical microprocessor modules, complete with memories and peripheral, all connected via a bus interface to a global bus. On the global bus, apart from the computer modules, are connected shared resources, such as memory banks and special memories acting as synchronization semaphores. Each microprocessor shares the same address space and each microprocessor can access the shared resources, such as the common memory or the local memory of another module, via the global bus. The local memories of any computer module are directly accessible from any other microprocessor, and all of the local memories form the common memory of the system.

Some computer modules are developed with an inner private bus containing strictly private resources not accessible by other processors. Figure 1.15 shows the μ^* architecture; in the figure the two solutions of microprocessor modules are represented. The first solution refers to a completely accessible local memory, the second one represents a two level hierarchy, where local memory is divided in two portions: one strictly private to the owner microprocessor and one local, but accessible, from the other microprocessors.

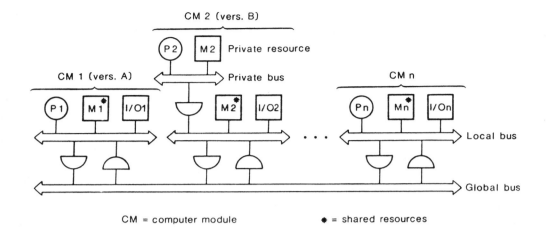

Fig. 1.15 – MSBI description of the μ^* multiprocessor system

1.4.6. The iAPX432 System

The iAPX 432 system developed by Intel represents a new design
approach to multiprocressor development,   based on VLSI technology
|IAPX81|. The iAPX432 is a family of integrated circuits that can be
freely combined to easily achieve complex multiprocessor systems.
Multiprocessor systems can now be performed at chip level rather
than at board level as described before.   One of the most important
objectives of the 432 system was to match the expanding needs with
increasing capability of a modular multiprocessor oriented system.
        The 432 hardware organization is composed of three different
types of modules:

    1) the interface processor (IP);
    2) the generalized data processor (GDP);
    3) the memory controller and the associated memory.

These three types of modules are connected via an intermodule
communication channel called 'packet bus'.   Each module is divided
into two logical parts: the processing (or the memory) element, and
the Bus Interface Unit (BIU).   The processing elements or the
memory element are connected to the local bus,   while the BIU acts
as connecting device toward the packet bus.

The interface processors and the generalized data processors can use one or more BIU to interface one or more packet bus. The GDP module is the central processing unit of the 432, more than one GDP module can be present in a system. The IP modules provide the interface between the external world and the 432 system, they are used to manage all the I/O traffic and to provide a protected interface for the GDPs. The memory modules are composed of the memory array and a memory controller unit (MCU) that interfaces the memory into the packet bus. The memory with its MCU can be connected to only one packet bus, this means that it is not possible to share the memory on more buses, or with more MCU like the BIU mentioned before. The packet bus is a multiprocessor message based communication channel of the 432 system, more than one packet bus can be used, depending on the application. As is now apparent, the 432 system can be expanded simply by replicating VLSI components or replicating functional blocks (IP+BIU, GDP+BIU, MEM+MCU) in order to increase computing power, as shown in Figure 1.16. Replication can also be used to increase reliability, for instance functional blocks can be doubled putting one block in 'master' mode, and the second in 'checker' mode. In this case any operation performed is considered correct only if the 'master' results match the results produced by the 'checker'.

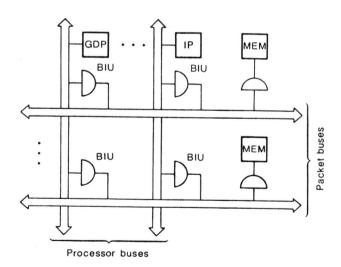

Fig. 1.16 – MSBI description of an iAPX432 multiprocessor system

1.4.7. The TOMP Multiprocessor System

The main objective of the TOMP research project concerned the design of a general frame under which a large set of multiple processor systems can be built up.

The architecture is needed to be able to support implementations within a wide range of interconnection strategies and computing power. The basic constraint was the use of commercially available microprocessors, without, however involving a strict dependance on just one of them. The system had to be defined and designed in such a way to allow the use of the available 16 bit processors, and also of the new ones in the same range. For this reason the work followed a strictly top-down approach: the functions to be supported by the system were specified in the first design step and, following these guidelines, the system structure was refined in successive steps.

Among the whole set of function needed in a multiprocessor environment only a limited set was chosen as a specific goal of investigation:

- processor/processor communication techniques;
- protection in multiple processor environment;
- modularity and structured design.

The next step was to design and realize, as a research prototype, one of the possible multiprocessor systems. The prototype work consists of the specification and the functional design of the system, and the implementation and the test of some modules. Some interconnection structures that had been considered for TOMP architecture are shown in Figure 1.17. All of them allow shared memory space, and use a system-wide bus to support the interprocessor communications.

Memories and buses of the structures shown in Figure 1.17 can be assigned to different categories, depending on access rights:

PRIVATE
The resource can be accessed only by the masters directly connected to it, that is belonging to the same processor.

LOCAL
The resource can be accessed directly by masters of the same processor module and, through the Global Bus, by masters of other processor modules.

GLOBAL
All access opportunities for all masters of all processors are equal.

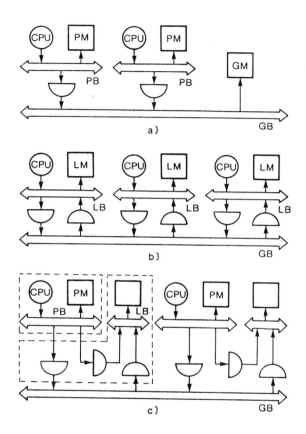

Fig. 1.17 – Bus-based architectures with shared memory
    a) Single global memory
    b) Private memories partially accessible from outside
    c) Dual-port memories used as shared resources

Single fully-global bus architectures, where no private or local resource is allowed are not shown because with this connection scheme the bus saturates even with few processors. A comparison of the various connection structures can (see chapter 5) show that architectures based on dual-port memories give better performance under different workloads.

The system is organized around two basic modules that allow the set-up of different communication structures. As shown in Figure 1.17c, they are respectively a processor board, and a board containing the dual-port memory and the local-to-global bus interface.

The processor boards were developed around the Zilog Z8001 microprocessor. Since these CPU boards contain local memory and parallel/serial I/O, they can also work autonomously as single-board computers.

The memory/bus interface board contains up to 16 kword of dual-port RAM, and a bus interface which supports the features provided on the global bus for interprocessor communications.

To ease the design of modules and interfaces, maximum compatibility between all bus levels has been enforced. Any bus which is brougth to a board connector must comply with the same specification. This allows the use of the same design for modules tied to the global bus or to the private bus of a processor.

An example of the TOMP system is described using MSBI notations in Figure 1.18. The same basic modules (CPU and dual-port memory) can be connected in other configurations and allow the set up of all the multiprocessor architectures shown in Figure 1.17.

One of the main features of the TOMP architectures is the possibility of functional growth by the addition of new modules. This kind of enhancement consists in the addition of primitive functions, such has a hardware management of task allocation, or memory protection.

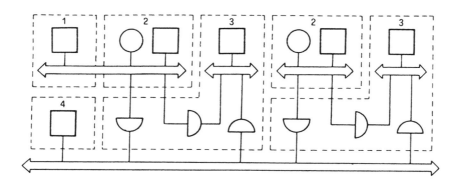

Fig. 1.18 - Physical modules and TOMP architectures:
a two- processor TOMP multiprocessor system

Usually the addition of a new function to a system needs the redesign of some module. A different approach is to associate the introduction of a set of new functions to the addition of a new module. This is possible if the system is carefully designed and great attention is devoted to the level or section at which insertion and addition is allowed.

In the TOMP architectures this level is the system bus. Some basic mechanisms must be designed into the protocol supporting the information exchange performed on the bus, in order to allow the new modules to enter into the communication dialogue performed by the modules already present in the system. Furthermore, the availability of these mechanisms allows us to keep undefined the complete set of functions that can be performed by the system. The only definition could be: "all the functions allowed by the basic mechanism can be performed using the proper module".

The system bus at which the insertion is possible is the M3BUS, which has been designed to fulfill the above mentioned goals. M3BUS |M3BU81| is described in chapter 5. The modules which exploit this feature and perform an added-on function will be here called SUPERVISORS. The functions that can be added at this level are all that can be done using just the information contained in the dialogue among modules, that is addresses, data and type of operation on data. The operations allowed are for example:

- memory protection: an operation is made or aborted according to the memory area involved;
- memory management: it basically involves the mapping of logical addresses output by a module into physical addresses calculated following some rule.
- implementation of a capability system.

It will be shown in chapter 4 how these rather complex functions can be performed only allowing a module to enter the information exchange between memory and processors. Obviously this new function must be conveniently supported by the upper layers of the system.

1.5 REFERENCES

|BARB81| Barbacci, M.R., "Instruction Set Processor Specifications
             (ISPS): The Notation and its Applications", IEEE Trans.
             on Computers, January 1981.

|BELL71| Bell, C.G., Newell A., "Computer Structures: Readings
             and Examples", McGraw-Hill, New York, 1971.

|BOWE80| Bowen B.A., and Buhr R.J.A., "The Logical Design of
             Multiple-microprocessor   systems",   Prentice-Hall,
             Englewoods Cliffs (N.J.), 1980.

|BRAD79| Brad W.H., "The Design and Implementation of a PMS
             Level Hardware Interconnection Language", Carnegie
             Mellon University, October 1979.

|CIVE82| Civera P., Conte G., Del Corso D., Gregoretti F., and
             Pasero E.,   "The μ^* Project: An experience with a
             Multimicroprocessor System", IEEE Micro, May 1982.

|ENSL78| Enslow P.H., "What is a Distributed Data Processing
             System", IEEE Computer, January 1978.

|FENG81| Feng T.Y., "A Survey on Interconnection Networks",
             IEEE Computer, December 1981.

|FLYN72| Flynn M.J., "Some Computer Organizations and their
             Effectiveness", IEEE Trans. on Computers, September
             1972.

|HWAN84| Hwang K., and Briggs F.A., "Computer Architecture
             and Parallel Processing", McGraw-Hill, New York, 1984.

|IAPX81| Tyner P., "iAPX432 General Data Processor Architecture
             reference manual", Intel Corporation, January 1981.

|JONE80| Jones A.K.   and Schwarz P., "Experiences Using
             Multiprocessor Systems – A Status Report", ACM
             Computing Surveys, June 1980.

|KATS78| Katsuk B.,   et al.,   "PLURIBUS– An operational fault-
              tolerant   multiprocessor",   Proceedings   IEEE,   October
              1978.

|M3BU81| Del Corso D.,   and Duchi G.,   "M3BUS: System specifi-
              cation   for   high   performance   multimicroprocessor
              machines", BIAS 1981 Proc., Milano, October 1981.

|MEAD80| Mead   C.,   and   Conway   L.,"Introduction   to   VLSI
              systems", Addison–Wesley, Reading (Mass.), 1980.

|MCFA82| McFarling   S.,   Turney   J.,   and   Mudge   T.,   "VLSI
              Crossbar Design Version Two",   CRL–TR–8–82, University
              of Michigan, February 1982.

|PARK83| Parker   Y.,   "Multi–microprocessor   systems",   Academic
              Press, London, 1983.

|SWAN76| Swan R.J.,   Fuller S.H.,   and Siewiorek D.P.,   "Cm*: a
              modular   multimicroprocessor",   Carnegie   Mellon   Univer-
              sity, November 1976.

|WULF72| Wulf   W.A.,   and   Bell   C.G.,   "C.mmp   –   A   Multi-
              miniprocessor",   Proc.   AFIPS   fall   Joint   Computer   Conf.,
              N.J. 1972.

CHAPTER 2

PERFORMANCE ANALYSIS OF MULTIPROCESSOR SYSTEMS

M.Ajmone Marsan, G.Conte
Dipartimento di Elettronica
Politecnico di Torino

G.Balbo
Dipartimento di Informatica
Universita' di Torino

ABSTRACT   This chapter describes a multiprocessor performance evaluation case study, and the modeling tools that were developed for this purpose. It is shown how the choice of the architecture of a multiprocessor system can be guided by analytical performance predictions in conjunction with implementation issues.   The goal of the project of a multiprocessor system is the development of an efficient architecture which should not experience bottlenecks at the physical level due to contention for shared resources.   A description is given of how the comparison among candidate architectures must be done using common assumptions and a similar workload model. Only after this preliminary work can an architecture be chosen and implemented as a good compromise between performance and implementation costs.

2.1. PERFORMANCE EVALUATION OF BUS ORIENTED MULTIPROCESSOR SYSTEMS

2.1.1. Introduction

The advantages of multiple processor systems over high speed and high power monoprocessor computers can be exploited if two basic conditions are met:

1. the computational problem is decomposed according to the distributed nature of the multiprocessor computer to profit from the parallelism of the system;

2. system overhead due to processor cooperation is kept low.

G. Conte and D. Del Corso (eds.), Multi-Microprocessor Systems for Real-Time Applications, 33–86
© 1985 by D. Reidel Publishing Company.

Focusing our attention on system overhead,   we can identify two
factors that contribute to the reduction of the overall system
efficiency:

  1. processor cooperation is managed by an executive program
      that uses processing power doing no "useful" work;

  2. contention for use of a limited number of common resources
      may cause processors to queue,   so that time is lost
      waiting.

Multiprocessor systems like C.mmp |WULF72| or Cm* |SWAN77| were
studied in the past to assess their efficiency.   Several authors
developed models to study the performance degradation of these
systems due to memory interference.   Some of these studies model the
system at the instruction execution level;   instructions are thus
represented as synchronous operations.   In other cases a more
abstract view of the system is used and asynchronous operations
are assumed that correspond to the execution of program segments
and to memory accesses for transfering variable quantities of data
(exponential assumptions are often introduced to simplify the
analysis of these models).   All of these studies assume that
processors and common memory modules are connected by a crossbar
switch.   Significant references in this field are |BASK76,   BHAN75,
HOOG77, SETH77|.
      Crossbar switches for the connection of many processors and
many memory modules are becoming less and less interesting due to
their complexity and their high cost (compared to the decreasing
costs of both processors and memories).   Moreover,   the bandwidth
provided by such interconnection structures often exceeds the
application requirements.   Recent proposals and implementations
indicate that bus structured interconnection networks are best suited
to multimicroprocessor systems |LEVY78,   THUR72, KAIS80|. With this
approach many different solutions for the interconnection network
are possible,   depending on the location of the shared memory
modules and on the structure of the processing units,   but little is
known about the efficiency of each alternative.   Hoener and Roeder
|HOEN77| presented a simple probabilistic analysis of bus contention
in a single bus multiprocessor system where processors are
organized in a priority hierarchy.   Willis |WILL78| considered a
simplified model of multiple bus systems,   assuming no queueing for
busy resources.   Fung and Torng |FUNG79| developed a deterministic
tool for the analysis of memory contention and bus conflicts in
multiple bus multiprocessor systems.   Ajmone Marsan and Gregoretti

|AJMO81| used an asynchronous model to analyze the performance of a single bus multiprocessor system with a single common memory module. This analysis was extended to multiple bus and multiple common memory systems by Ajmone Marsan and Gerla |AJMO82a|.

In this chapter we describe the performance evaluation studies that were performed in the early stages of the development of TOMP |CONT81|.

Several candidate architectures were studied, and their performances were compared using common assumptions and the same workload model. Based on this preliminary work one architecture was choosen and implemented. In a subsequent step the analysis of the selected architecture was broadened using new modeling tools that allow us to explore the performance of some extensions of the architecture chosen for TOMP.

We first introduce simple analytical models for the performance analysis and comparison of several single bus multiprocessor architectures. As in the case of some of the crossbar architecture studies, our models are derived using a somewhat abstract point of view. Processors are assumed to execute programs stored in their own private memories. The execution of a program is interleaved with accesses to common memories. One can thus identify processor bursts alternating with transfer periods required to move variable quantities of data from common memories to private memories and vice versa. Contention may arise for the use of the global bus connecting processors and memories, and for accessing common memory modules.

Four architectures that differ for the location of the common memory modules are studied using simple Markovian models of their behaviour. Quantitative results are obtained, which show that, in fairly large systems, the performances of three of the four architectures are very similar, whereas the other one behaves much worse.

The choice of one of the four architectures is based on performance estimates and implementation issues. The analysis of the chosen architecture is then extended by means of new tools such as Stochastic Petri Nets and Queueing Networks. This makes possible to investigate the impact of architectural extensions on system performance, and to predict the changes in efficiency that can be obtained.

Finally, in order to validate the analytical performance predictions obtained with the stochastic models, measurements on the actual system are reported. Actual data show that measurement results are in good agreement with the performance estimates.

This chapter is organized as follows: Section 2.1.2 introduces

the four architectures considered in this study and describes the
general assumptions used in the construction of their Markov
models. Further information about the level of detail used in the
models and about the system workload are presented in Section
2.1.3. Sections 2.1.4 to 2.1.7 present the particular assumptions
used for each architecture and illustrate their model construction
and analysis. In Section 2.1.8 numerical results are shown and the
four architectures are compared; the choice of the architecture of
TOMP is then discussed in Section 2.1.9.

Section 2.2.1 introduces the tools used to extend the analysis
of the chosen architecture, and describes their characteristics.
Sections 2.2.2 and 2.2.3 are devoted to the analysis of extension of
the TOMP architecture using Stochastic Petri Nets and Queueing
Networks, respectively.

Section 2.2.4 reports on the measurements performed on the
TOMP prototype available at Politecnico of Torino. Finally,
measurement results are shown to be in very good agreement with
the anlytical performance predictions.

2.1.2. Modeling Assumptions

In this study we consider four single bus multiprocessor
architectures in which processor modules include a CPU and a
memory. Other memory modules, not directly tied to any processor,
may also be available.

From a physical point of view we can thus recognize, with
respect to a given processor, local and external memory modules. A
processor accesses its own local memory module with a single level
bus connection, and external memory modules with a multiple level
bus connection. There may be memory modules which are not
reachable from a given processor.

From a logical point of view we can identify private and
common memories (again with respect to a given processor). Private
memories are accessible only from the processor to which they are
local. Common memory modules are accessible from all processors.

We present in this section the assumptions, common to the four
architectures, that are used to obtain simple models of the system
behaviour.

Processors are assumed to execute a continous flow of
instructions stored in their own private memories. These instructions
are logically grouped in tasks that cooperate by passing messages
through common memory areas. The common memory can be
implemented either using one (or more) memory module(s) external to

all processors,   or distributing it in the non private part of the local memories.

Four different common memory organizations are analyzed and compared in the following, details will be given separately for each architecture.

The execution of tasks cooperating in a message passing fashion therefore amounts to repeatedly executing a CPU_burst – transfer_period cycle.   Contention for the use of shared resources, such as global bus and common memory modules,   can add a queueing period component to the execution cycle. Our assumption of a continuous flow of instructions being executed by each processor implies that idle periods due to task synchronization are negligible or,   equivalently,   that the number of tasks allocated to processors is very large with respect to the number of processors itself.

We can thus classify the state of a processor as follows:

1. ACTIVE. The processor executes in its private memory.

2. ACCESSING.   The processor exchanges information with other cooperating processors by writing into (or reading from) common memory areas.

3. QUEUED.   The processor queues waiting to access common memory areas.

4. BLOCKED.   The processor is blocked by some other processor accessing the common memory segment of its local memory.

Parameters of our models are the average CPU burst length $(1/\mu)$ and the average tranfer period duration $(1/\lambda)$.

The performance index used here is the average number of active processors, called processing power and denoted by P.

Many other performance indices can be derived from P as shown in |AJMO82a|. When presenting results we sometimes normalize P for the number of processors in order to allow a better comparison between systems with a different number of processors.

2.1.3. The system workload

In order to compare the different multiprocessor systems,   it is necessary to define a workload which is independent on the architecture.   Processors are then assumed to execute tasks that can communicate either with other tasks allocated to the same processor,

or with tasks allocated to other processors.   Associated with each
task is an "input port" stored in the private memory of the
processor to which the task is allocated.   Every message issued by
the task is directed to the input port of the destination task.
Communication between tasks allocated to the same processor takes
place through private memory only.   No common resources (common
memory and global bus) are involved in the message exchange
operation.   Communication between tasks allocated to different
processors must use a "communication port" residing on common
memory (which may be local to processors).   One communication port
is associated with each processor as its input port.   A pictorial
view of the logical structure of the communication between tasks is
shown in Figure 2.1.

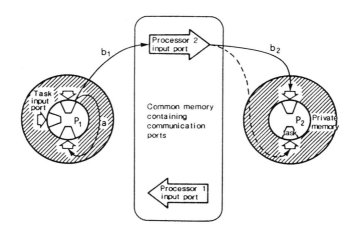

Fig. 2.1 – Logical structure of the comunication between tasks:
        a) a task allocated to P1 puts a message into the
           input port of another task allocated to the same
           processor;
        b) a task allocated to P1 sends a message to a task
           allocated to P2 with a two steps action:
           b1) the message is put into the input port of P2,
           b2) the message is moved to the input port of the
           destination task.

Transfer period durations always depend on the length of the
message being written in (read from) the external common memory
and are thus independent on the system architecture.   CPU bursts,
on the other hand, may have different average lengths in different

architectures. Depending on the organization of the message passing mechanism, CPU bursts may comprise accesses to local common memories needed to complete the delivery of the message to the destination task.

In order to produce a useful piece of information (a message), a task must execute for a random time with average $1/\lambda_t$.

The exchange of information between tasks is relevant to our models only when the communicating tasks reside on different processors since only in this case the shared resources are used. The parameter we need, in order to study the contention for shared resources, is the rate of generation of messages exchanged among processors. We define $1/\lambda_p$ to be the average active time elapsing between subsequent messages sent out by the same processor towards communication ports.

A relationship between λ_t and λ_p is easy to obtain within the framework of a global workload of n tasks. Assume that the number of tasks (n) is a multiple of the number of processors (p) considered in the model, and that exactly n/p tasks are allocated to each processor. The number of tasks external to each processor is simply $n-(n/p)$. If we assume a uniform reference model between tasks, the probability that task i sends a message to the input port of task j is $1/(n-1)$ for all $j \neq i$. Thus we have:

$$(2.1) \qquad \lambda_p = \lambda_t \frac{(n-n/p)}{(n-1)} = \lambda_t \frac{n(p-1)}{p(n-1)}$$

Moreover, if we assume n to be very large, we can approximate λ_p as:

$$(2.2) \qquad \lambda_p = \lambda_t \frac{(p-1)}{p}$$

The parameters λ_p and λ_t are both useful: λ_p determines the behaviour of the different models once the number of processors is fixed. λ_t is needed in order to compare the performance of each architecture when a different number of processors is used to execute the same (fixed) workload.

A relationship between λ and λ_p can be derived from the operating characteristics of each architecture and considering the two step interprocessor communication scheme shown in Figure 2.1. The details of this derivation will be provided with the discussion of the models used to represent each architecture.

Results are given in terms of ϱ, ϱ_p, and ϱ_t, that are obtained as the ratio of λ, λ_p, and λ_t to μ. The above quantities do not represent traffic intensities but can be interpreted as the characterization of the workload seen by the model (ϱ), generated by a processor (ϱ_p), and generated by a task (ϱ_t).

For our models to be "simple" and computationally not as expensive as either simulation or prototype setups, it is necessary to assume that the underlying stochastic processes satisfy the Markov property. To this purpose the following additional assumptions are introduced:

1. The durations of CPU bursts and access periods are assumed to be exponentially distributed random variables.

2. When a processor requires access to a common memory module, a path is immediately established (with no delay) between the processor and the referenced memory module, provided that the memory is accessible (the bus is available and the memory is free).

3. If a path cannot be established, the processor idles, waiting for the necessary resource(s).

4. Upon memory access completion, memory and busses are immediately released (with no delay) and the processor returns to its active state.

5. An external access request from processor i is directed to a non local memory j with probability 1/m, where m is the number of non local common memory modules.

Assumptions 2 and 4 imply that we neglect the times needed for the bus arbitration and release. This is a consequence of our rather abstract view of the system behaviour. Moreover it seems reasonable to assume that these times are of at least one order of magnitude smaller than those associated with CPU activity and data transfering. Nothing however prevents us from adding both arbitration and release time to the access time in the period of average duration $1/\mu$ used in the model.

To further simplify our models we considered completely symmetric systems. While this restriction is not necessary to preserve the Markovian properity of the models, its introduction allows a substantial reduction of the complexity of their analysis. The impact of the symmetry assumptions on the quality of the results is discussed for a similar case in |AJMO82a|.

2.1.4. Architecture 1

The first architecture we consider is characterized by the existence of a common memory external to all processors and accessible only through the global bus. This memory contains all the communication ports. Contention arises each time a message is written in (read from) common memory. Only one processor can access common memory at each point in time. Figure 2.2 depicts a 3 processor system organized according to this structure.

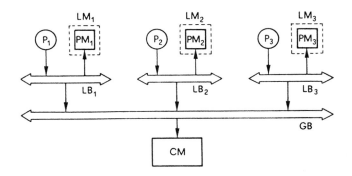

Fig. 2.2 – Structure of architecture 1 with single common memory and 3 processors composed of CPU (P_i) and private memory (PM_i) connected by a local bus (LB_i).

The behaviour of this system at the message passing level can be described as follows: processors execute segments of programs until they need to access common memory. After issuing a request for the global bus, processors may have to wait for the bus to become available. When the bus is allocated to a processor the transfer of data takes place. In a message passing environment each message sent out by a processor is eventually read by the destination processor. The symmetry of the model implies that (on the average) the traffic of messages flowing out of a processor balances that flowing into the same processor. Processor activity is thus interrupted with a rate (λ) that is twice the rate of generation of messages:

(2.3) $$\lambda = 2\lambda_p$$

The independency, and the exponential assumptions introduced in the previous section, make this model behave like a "machine repairman" with exponential service times (Palm's model |PALM58|). Solutions of this model are found in most queueing theory books, as it can be viewed either as an M/M/1/ /p queue, or as a central server queueing network (see for instance |KLEI75|).

The definition of processing power allows the derivation of the following closed form expression:

$$
(2.4) \qquad P = \cfrac{\displaystyle\sum_{k=0}^{p} \varrho^{k}\, \frac{p!}{(p-k)!} - 1}{\varrho \displaystyle\sum_{k=0}^{p} \varrho^{k}\, \frac{p!}{(p-k)!}}
$$

where p is the number of processors and ϱ is the load factor. A recursive formula can also be given:

$$
(2.5) \qquad P(p) = \frac{p}{1 + \varrho\;\; p - P(p-1)}
$$

Using the relationship between ϱ, ϱ_p, and ϱ_t:

$$
(2.6) \qquad \varrho = 2\,\varrho_p = 2\,\varrho_t\; \frac{(p-1)}{p}
$$

equations (2.4) and (2.5) can be written in a form that will be later used for comparison purposes.

2.1.5. Architecture 2

As mentioned in Section 2.1.1, the common memory can be distributed on modules local to each processor. Architecture 2 assumes that local memories are (logically) divided into private and common areas. Common areas contain the communication ports of the associated processors. Each processor is connected to its own local memory segment by a local bus. A processor accesses a non local common memory segment using its own local bus, the global bus, and the local bus connected to the destination common memory module where the input port of the destination processor is located.

Figure 2.3 depicts a 3 processor system organized according to this structure.   Arbitration mechanisms are needed to manage global and local busses. Contention may arise for using each of the busses represented in the model.

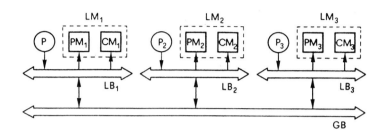

Fig. 2.3 – Structure of architecture 2 with distributed common memory (CM_i) and 3 processors composed of CPU (P_i) and associated private memory (PM_i).

A processor that gains access to the global bus acquires priority to use any reachable resource and may preempt other processors. Processors preempted while active become blocked; processors preempted while queued maintain their state, but release their local bus. These policies avoid deadlocks and improve performance.

In this architecture, a message generated by a task executing on the sender processor (processor i) is passed to the input port of a task allocated to the destination processor (processor j, j ≠ i) using the following mechanism: at the end of a CPU burst, processor i issues a request for the global bus. When available, the bus is seized by the processor together with the local bus j: a transfer period begins and data are moved into the input port of processor j.   The message is eventually received when the destination processor moves it from its input port to the task input port.   This latter action is a transfer of data within the local memory of the destination processor and is thus considered part of its activity, but must not be regarded as a contribution to the processing power.

As we observed in the previous section, the symmetry of the model implies that flows of incoming and outgoing messages balance. It follows that the mean length of a CPU burst can be considered as the sum of the mean time required to create a message $(1/\lambda_R)$ and the mean transfer period required to receive a message $(1/\mu_R)$. Hence:

(2.7) $1/\lambda = 1/\lambda_p + 1/\mu$ or $\lambda = \lambda_p \mu/(\lambda_p + \mu)$

Because of the blocking phenomenon due to one processor accessing the local memory of another one, architecture 2 cannot be modeled directly as a simple queueing system. A Markov chain model can nevertheless be constructed, provided that the system state description is correctly chosen. In the case of architecture 2 the state of the system is defined by the 2p-tuple:

(2.8) $(m_1, s_1 ; m_2, s_2; \ldots ; m_p, s_p)$

where m_i is the index of the memory module referenced by processor i and s_i is the state of processor i.

s_i can take the values:
      2 active
      1 accessing an external common memory module
      0 blocked
      -k queued for the global bus: k-th in queue.

The symmetry of the system can be used in conjunction with the theory of "lumpable" Markov chains |KEME60| in order to reduce the number of states of the chain. Aggregated states require a less detailed description. The state definition in the lumped chain is given by the triplet:

(2.9) (n_a , n_e , n_b)

where:

      n_a = number of active processors,

      n_e = number of processors either accessing an external common
            memory area, or queued for the global bus,

      n_b = number of blocked processors (which were active and
            have been preempted by an external access).

An important property of the lumped model is its size, which grows only linearly with the number of processors in the system: the analysis of systems in which a very large number of processors cooperate is thus feasible. The state transition rate diagram of the lumped Markov chain for the general case of p processors is shown in Figure 2.4, and the expressions for the transition rates are given in Table 2.1.

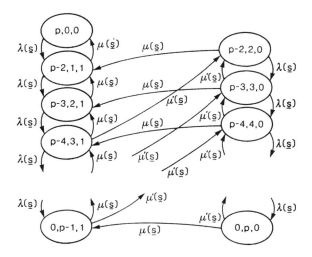

Fig. 2.4 – State transition rate diagram for the lumped Markov chain of architecture 2.

Notice that λ-transitions correspond to the generation of an access request by one of the active processors. μ-transitions correspond to the completion of common memory accesses.
    Processors requesting the use of the global bus when it is busy,   wait in the queue for their turn.   Upon completion of an access   to   the   external   common   memory,   the   global   bus   is immediately seized by one of the queued processors.   A new message transfer begins and the corresponding destination processor becomes blocked   unless   it   was   already   waiting   in   the   global   bus   queue. Depending on the state of this target processor,   the system moves into   a   state   with   either   one   or   zero   blocked   processors.   Associated with the first alternative are μ-type transitions while μ'-type transitions correspond to the second one. Given a state $\underline{s}=(n_a, n_e, n_b)$ the access request generation rate is proportional to the number of active   processors n_a;   the   total   access   completion   rate   is μ whenever n_e is larger than zero.   If n_e is less then or equal to 2 only μ-type transitions may take place because when the accessing processor completes its access at most one processor is queued for the global bus,   and it will surely block an active processor. If n_e is larger than two then two cases are possible: upon the end of an access   a   processor   seizes   the   bus   and   may   either   access   the   local

Table 2.1

Transition rates of the lumped Markov chain of architecture 2 (Figure 2.4)

$$\lambda(s) = n_a \lambda$$

$$\mu(\underline{s}) = \begin{cases} \left(1 - \dfrac{n_e - 2}{p - 1}\right) \mu & n_e > 2 \\ \\ \mu & n_e \leqslant 2 \end{cases}$$

$$\mu'(\underline{s}) = \dfrac{n_e - 2}{p - 1} \mu \qquad\qquad n_e > 2$$

where $\underline{s} = (n_a, n_e, n_b)$

note that $\mu(\underline{s}) + \mu'(\underline{s}) = \mu$

$$n_a + n_e + n_b = p$$

---

memory of one of the n_e-2 processors queued for the global bus or the local memory of an active processor. The first alternative corresponds to a μ'-type transition and due to the uniform probability distribution, may occur with probability $(n_e-2)/(p-1)$. The second alternative corresponds to a μ-type transition and may occur with probability $1-(n_e-2)/(p-1)$.

The equilibrium probabilities of the states of the lumped chain are easily evaluated by solving a system of linear equations. Due to the regularity of the structure of the lumped Markov chain, it is not difficult to set up a program that automatically generates the states of the chain and evaluates the equilibrium probabilities for systems of any reasonable size. Computational problems arise for very large systems (hundreds of processors) in the solution of the system of linear equations that gives the equilibrium probabilities.

Let S be the state space of the Markov chain, s be a state and $\pi(s)$ be its equilibrium state probability. The processing power of the multiprocessor system is given by the following expression:

$$(2.10) \qquad P = (1 - \varrho) \sum_{s \subset S} n_a(s) \, \pi(s)$$

The factor $(1 - \varrho)$ is introduced to account for the message read time included in the CPU bursts as expressed in (2.7). The fraction of CPU burst actually used to generate a message is indeed:

$$(2.11) \qquad \frac{\dfrac{1}{\lambda} - \dfrac{1}{\mu}}{\dfrac{1}{\lambda}} = 1 - \varrho$$

Closed form expressions for the processing power of this architecture with two, three and four processors are given in Table 2.II, as functions of ϱ.

Table 2.II

Architecture 2, processing power of multiprocessor systems with two, three, and four processors.

NUMBER OF PROCESSORS	PROCESSING POWER
2	$\dfrac{2(1 - \varrho)}{1 + 2\varrho}$
3	$\dfrac{3(1 - \varrho^2)}{1 + 3\varrho + 3\varrho^2}$
4	$\dfrac{4(1 - \varrho)(9 + 36\varrho + 29\varrho^2 + 52\varrho^3 + 24\varrho^4)}{9 + 54\varrho + 149\varrho^2 + 236\varrho^3 + 208\varrho^4 + 96\varrho^5}$

Substituing for ϱ its definitions in terms of ϱ_p and ϱ_t:

(2.12) $$\varrho = \frac{\varrho_p}{1 + \varrho_p} = \frac{\varrho_t}{\varrho_t + \frac{p}{p-1}}$$

the equations of Table 2.II become useful for comparison purposes.

2.1.6. Architecture 3

An improvement on architecture 2 can be obtained using a double-
port memory module |CHAN80| to implement the common part of the
local memory of each processor.   Common memory modules are thus
directly accessible from external processors through the global bus.
        No contention arises either on local busses or on double port
memories that support two simultaneous accesses.   Contention is in
this case only due to the sharing of the global bus.   Figure 2.5
depicts the structure of architecture 3 in the case of a two
processor system.   Note that processors are not allowed to access the
common section of their own local memory through the global bus.

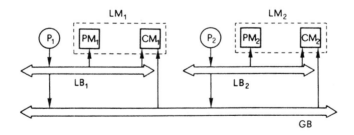

Fig. 2.5 – Structure of architecture 3 with a distributed common
           memory (CM_i) and 2 processors composed of CPU (P_i) and
           local memory (LM_i) conected by a local bus (LB_i).

Passing messages between tasks allocated to different processors
involves a sequence of operations very much similar to that
described in the previous section.   Because of the double port
memory, writing a message in the input port of the destination
processor does not block its activity.   Again CPU bursts include the
action of moving a message from the processor input port to the

task input port, both located in the processor local memory, so that the relationship between λ and λ_p remains the one described by equation (2.7) that we repeat here for convenience:

$$(2.13) \qquad \lambda = \frac{\lambda_p \mu}{\lambda_p + \mu}$$

Since the global bus is the only element that may cause contention, the Markovian model of architecture 3 is again a "machine repairman" model, and we can use the results of Section 2.1.4 for the evaluation of this architecture.

The expression of the processing power obtained for architecture 1 can be used, but, like in the case of architecture 2, it must be reduced by a factor $(1 - \varrho)$ that accounts for the time needed to transfer a message within the local memory between processor and task input ports. We thus obtain:

$$(2.14) \qquad P = \frac{(1 - \varrho) \; \displaystyle\sum_{k=0}^{p} \varrho^k \; \dfrac{p!}{(p - k)!} - 1}{\displaystyle\sum_{k=0}^{p} \varrho^k \; \dfrac{p!}{(p - k)!}}$$

and the recursion becomes:

$$(2.15) \qquad P(p) = \frac{p(1 - \varrho)}{1 + \left(p - \dfrac{P(p - 1)}{1 - \varrho}\right)}$$

Substituting for ϱ its definition in terms of ϱ_p and ϱ_t:

$$(2.16) \qquad \varrho = \frac{\varrho_p}{\varrho_p + 1} = \frac{\varrho_t}{\varrho_t + \dfrac{p}{p - 1}}$$

Equations (2.14) and (2.15) become useful for comparison purposes.

2.1.7. Architecture 4

When a double port memory is not available, a variation of
architecture 3 can be obtained implementing the common memory
modules as shown in Figure 2.6, that depicts the structure of
architecture 4 in the case of two processors. Each common memory
module contains the input port of its associated processor.
Nevertheless common memory modules are external to all processors.

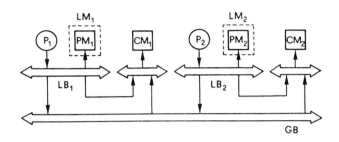

Fig. 2.6 – Structure of architecture 4 with distributed common memories (CM_i) and 2 processors composed of a CPU (P_i) and
private memory (PM_i) connected by a local bus (LB_i).

As in architecture 3, processors are not allowed to access their
associated common memory segments through the global bus. Only
one processor is allowed to access a common memory module at each
point in time. Contention arises for the use of the global bus and
of the common memory modules. Arbitration mechanisms are needed
for managing the global bus and the common memory busses. As in
architecture 2, to improve performance, priority is given to access
requests coming from the global bus: a processor accessing its
associated common memory module may thus be preempted.
        In this architecture a message generated by a task executing
on the sender processor (processor i) is passed to the input port of
a task allocated to the destination processor (processor j, j ≠ i)
using the following mechanism: at the end of a CPU burst,
processor i issues a request for the global bus. When available,
the bus is seized by the processor together with the bus of the
destination common memory module. A transfer period begins and
data are moved into the input port of processor j. During this
transfer period processor j is blocked if reading messages from its
input port. The message is eventually received when the destination
processor moves it from its input port to the task input port. In
this architecture (as in architecture 1) the latter action is

considered a transfer period since it involves an external memory access. Processor activity is thus interleaved with external common memory accesses to read and write messages. Because of the symmetric workload assumption, incoming and outgoing message flows balance. Processor activity is thus interrupted at a rate (λ) which is twice the rate of message generation:

(2.17) $\lambda = 2 \; \lambda_p$

As in architecture 2, the system is modeled by a continuous time Markov chain with the same state definition as in (2.8). The only difference is that now we must distinguish whether the processor accesses its associated common memory module or an external one. We will thus use $s_i = 3$ meaning processor i is active, and $s_i = 2$ to represent processor i accessing its associated common memory module.

The complete Markov chain for the two processor case has the state transition rate diagram shown in Figure 2.7.

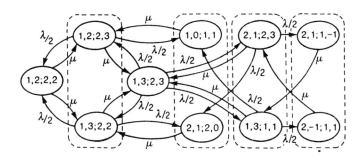

Fig. 2.7 – State transition rate diagram of the Markov chain for the 2 processor system, architecture 4.

Lumping the Markov chain corresponding to the general case is not as straightforward as for architecture 2. While we are only interested in first order performance indices related to the average number of active processors, a reduction of the state description, which takes into account the state of each processor and neglects the destination of the external reference, does not meet the theoretical conditions for the lumpability of the original Markov chain. The destination of the memory reference of processors waiting in queue as well as that of the processor working on external common memory are thus important. A direct application of the lumpability criteria to the transition probability matrix of the

Markov chain underlying the original model is computationally not
feasible due to the complexity of the problem,   even when a small
number of processors is considered.   The inherent symmetry of the
model can however be exploited to obtain a first reduction of the
state space size allowing a direct solution of slightly more complex
systems.

    This reduction, although exact, does not completely exploit the
power of the lumpability criteria.

    The state description we have chosen to perform this reduction
step has a structure similar to that of the original model.   The only
difference is that processors are ordered according to their activity
state (i.e.   active processors,   queued processors,   ... are grouped
together) without distinguishing among processor indices,   but
distinguishing among memory reference indices.

    The state definition used to describe the lumped model is thus
the following ordered list:

$$(2.18) \qquad (m_{1st}, s_{1st}; \; \cdots \; ; m_{i-th}, s_{i-th} \; ; \; \cdots \; ; m_{p-th}, s_{p-th})$$

where the position of each pair does not necessarily correspond to
the index of the associated processor.   The memory reference of the
i-th pair indicates the position held by the destination processor in
the ordered list.

    Using this state description a substantial reduction of the size
of the state space is achieved (see Table 2.III),   and an algorithm
can be devised to automatically generate the transition matrix of
the lumped Markov chain. Unfortunately, the size of the state space
of the lumped chain keeps growing combinatorially with the number
of processors considered in the model.   The exact computation of the
performance indices is thus feasible only for small models.

Table 2.III
Number of states in the Markov chain used to model architecture 4
vs. number of processors.

| Number of | Number of states | |
Processors	Original Chain	Lumped Chain
2	10	6
3	128	25
4	3784	173
5	---	1784

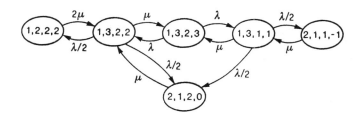

Fig. 2.8 – State transition rate diagram of the lumped Markov chain
for the 2 processor system, architecture 4.

Applying this reduction technique to the two processor chain of
Figure 2.7, aggregating the states comprised within dashed boxes,
we obtain the Markov chain of Figure 2.8, which can easily be
solved, giving a result for the processing power:

$$P = \frac{4(\varrho + 1)(\varrho + 2)}{3\varrho^3 + 11\varrho^2 + 10\varrho + 4}$$

(2.19)

As the number of system components increases, the analysis of the
model becomes more and more complex. We carried the analysis up
to a 5 processor system.

Comparison with other architectures is made possible by
substituting for ϱ its definition in terms of ϱ_p and ϱ_t:

$$\varrho = 2\,\varrho_p = 2\,\varrho_t\,\frac{p-1}{p}$$

(2.20)

2.1.8. Architecture Comparison

We start the architecture comparison by considering the two-
processor case, as it is the first step towards multiprocessing and
points out some results that become less obvious (but not less
important) in larger systems.

Figure 2.9 shows the processing efficiency as function of ϱ_p,
the processor communication load, for the four two processor
architectures. The same result are given in closed analytical form
in Table 2.IV. These results support the considerations used to
develop the four architectures as architecture 3 is superior to 4,

which, in turn, is better than 2. These considerations imply that architectures 2 and 3 provide, respectively, lower and upper bounds on the performance of architecture 4.

Table 2.IV
Processing power of two processor systems.

ARCHITECTURE	PROCESSING POWER
1	$\dfrac{2\ (\ 1\ +\ 2\ \varrho_p\ )}{1\ +\ 4\ \varrho_p\ +\ 8\varrho_p^2}$
2	$\dfrac{2}{1\ +\ 3\ \varrho_p}$
3	$\dfrac{2\ (\ 1\ +\ 2\ \varrho_p\ )}{1\ +\ 4\ \varrho_p\ +\ 5\ \varrho_p^2}$
4	$\dfrac{2\ (\ 1\ +\ 2\ \varrho_p\ )\ (\ 1\ +\ \varrho_p\ )}{1\ +\ 5\ \varrho_p\ +\ 11\ \varrho_p^2\ +\ 6\ \varrho_p^3}$

The mutual behaviour of architectures 1 and 2 is very interesting: for light loads, architecture 1 outperforms 2; this is rather surprising because architecture 1 generates twice as many accesses to the global bus. This result is due to the fact that, with light loads, the average queueing delay is very low, making thus negligible the additional contention introduced by architecture 1. With architecture 2, on the other hand, every access to an external common memory area preempts a processor, whose probability of being active on its local memory is very high under light load conditions. This same argument explains why architecture 2 is the only one for which the derivative of P with respect to ϱ_p is negative for $\varrho_p=0$. In all other cases we have a null derivative at $\varrho_p=0$. The break even point between architectures 1 and 2 is

ϱ_p =0.5;   for higher loads architecture 2 becomes advantageous. Low loads should,   however,   be considered as most significant for comparison purposes,   because well designed multiprocessor systems should operate in this region if the problem decomposition into tasks and the task allocation to processors is aimed to reducing communication overhead.

For very low loads we see that the behaviour of architectures 1, 3 and 4 are very similar. For two processor systems we can thus conclude that,   whenever the bus is not the system bottleneck, architecture 1 is not a bad choice, considering the simplicity of its implementation.

Considering now more complex systems,   we present in Figure 2.10 the processing efficiency of a five processor system organized according to the four different architectures;   Figure 2.11 shows the processing efficiency of a ten processor system.   In the latter case results are available only for architectures 1,   2 and 3,   but the result obtained for smaller systems support the conjecture that the performance curve of architecture 4 always lies between the curves of architectures 2 and 3.

The results provided by the two,   five,   and ten processor systems show some trends which allow making general statements about the behaviour of the different architectures.   The ranking of architectures made for the two–processor case remains valid also in more complex situations.   Increasing the number of processors,   the performances of architectures 2,   3,   and 4 become very similar, up to the point that differences become negligible in the case of the ten processor system even for very light loads.   Architecture 3 no longer gives a noticeable advantage as it did in the two processor case.   The similar behaviour of architectures 2,   3,   and 4 for heavily loaded large systems can be intuitively explained by the bottleneck effect of the global bus;   in these conditions processors are mainly queued for the global bus,   so that other contention and blocking phenomena tend to disappear.

A   further   consideration   is   that   the   crossover   between architectures 1,   and 2 now takes place for very low loads,   and, when the communication load increases,   architecture 1 behaves significantly worse than the others.

In Figure 2.12 we have plotted the processing power of architectures 1,   2, and 3 as a function of the number of processors in the system,   for a fixed processor communication load ϱ_p =0.1. The curve of architecture 4 is not shown, but again it lies between those of architectures 2 and 3.   The same considerations made for Figures 2.10,   and 2.11,   can be drawn from this figure too.   It provides strong evidence that architecture 1 is a bad choice for

large systems, and that the performance of architecture 4 needs not
be investigated further,   as architectures 2 and 3 provide tight
upper   and   lower   bounds.   Note   that   also   this   figure   shows   a
crossover between architectures 1,   and 2,   for increasing system
size.
        The   behavior   of   each   architecture   when   the   number   of
components is increased is shown in Figures 2.13-2.16.   Figures
2.13,   2.14,   2.15,   and 2.16 show,   respectively,   the normalized
processing power of architectures 1 to 4 versus ϱ_p for varying
system size.   It can be noted that architecture 1 shows the largest
performance reduction for increasing number of processors and
communication load.   Moreover,   for very low loads architectures 3
and 4 yield very similar performances,   superior to those of the
other two architectures.   This can be seen comparing the slopes of
the curves for low communication loads.   Denormalizing the results
for architecture 1 we observe that high communication loads induce
such   a   contention   for   the   global   bus   to   almost   nullify   the
advantage   expected   when   adding   new   processors.   The   same
phenomenon,   for   higher   loads,   is   observed   for   the   other
architectures too.   For all architectures these figures show how the
potential processing power provided by new processors translates
into actual processing power only when the communication load is
kept low.
        These results are obviously biased by the fact that the models
discussed   here   explicitly   neglect   performance   losses   due   to
synchronization   among   tasks   and/or   processors.   In   large
multiprocessor systems, our assumption of the processors executing a
continous flow of instructions,   that is the number of tasks being
much larger than the number of processors,   may not be justified
making these conclusions rather optimistic.
        All of the above results were derived using the assumptions
introduced in Sections 2.1.2 and 2.1.3,   that allowed us to obtain
simple Markovian models of the multiprocessor system behaviour.
Real systems do not quite possess those characteristics,   but it is
generally recognized that Markovian queueing models provide robust
performance estimates with respect to changes in the hypotheses (see
for example |BUZE77|).
        The hypothesis of exponentially distributed access times is
conservative with respect to all distributions with coefficient of
variation smaller than one;   in particular,   if the message duration
is   fixed,   or   can   vary   only   within   given   limits,   the   actual
performance will be better than predicted by our models.   The
approximation introduced by the exponential distribution assumptions
would be less satisfactory if the real system were characterized by

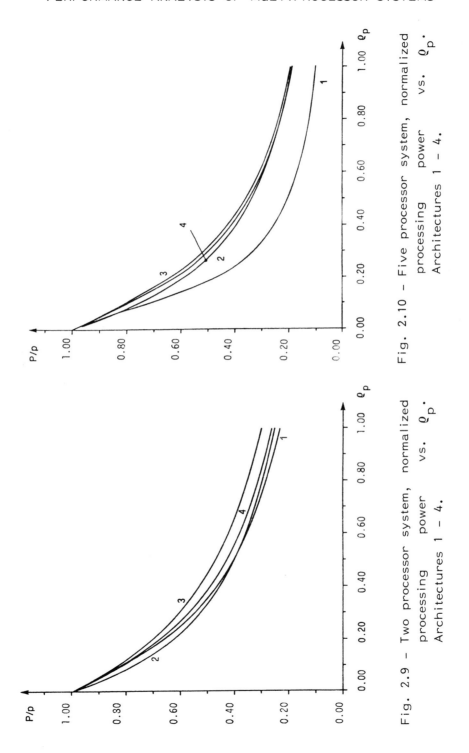

Fig. 2.10 – Five processor system, normalized processing power vs. ϱ_p. Architectures 1 – 4.

Fig. 2.9 – Two processor system, normalized processing power vs. ϱ_p. Architectures 1 – 4.

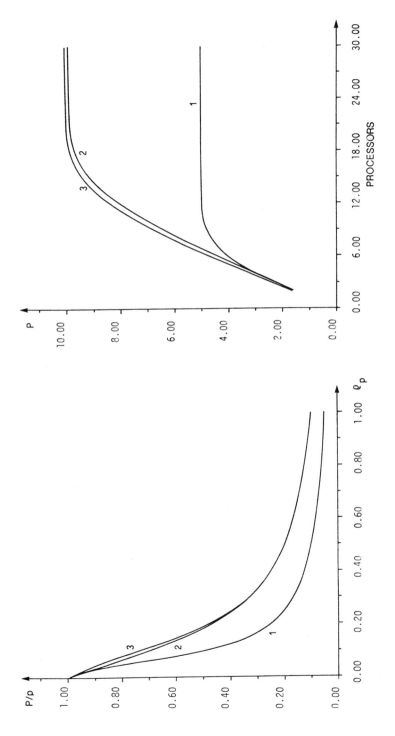

Fig. 2.12 – Processing power vs. number of
processors for ϱ_p = 0.1 in the
case of architectures 1, 2 and 3.

Fig. 2.11 – Ten processor system, normalized
processing   power   vs. ϱ_p.
Architectures 1 – 3.

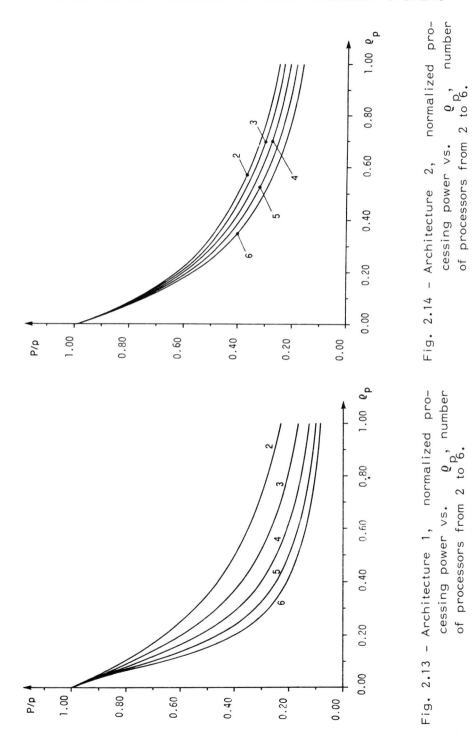

Fig. 2.14 – Architecture 2, normalized processing power vs. ϱ_p, number of processors from 2 to 6.

Fig. 2.13 – Architecture 1, normalized processing power vs. ϱ_p, number of processors from 2 to 6.

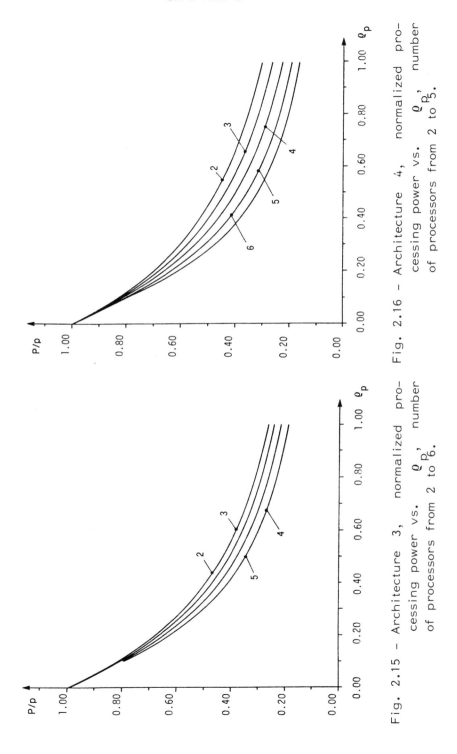

Fig. 2.16 – Architecture 4, normalized processing power vs. ϱ_p, number of processors from 2 to 5.

Fig. 2.15 – Architecture 3, normalized processing power vs. ϱ_p, number of processors from 2 to 6.

service time distributions with coefficient of variation greater than one; it should be noted that most of the resource allocation schemes implemented in real systems tend to keep this factor low, and hence tend to limit the approximation errors produced by these assumptions. The uniform distribution of load among processors is a conservative assumption too: since a processor cannot interfere with itself, the worst case is obtained when all processors generate messages with the same rate.

2.1.9. Choice of the Architecture of TOMP

Some qualitative considerations on the relative costs of implementation of the four architectures are possible: architecture 1 is the simplest and most inexpensive to implement with state of the art technology. The cost of architectures 2 and 4 are very similar, possibly being architecture 4 a little more expensive to implement, due to the need of multiport memories. Ruling out architecture 1 because of the poor performance in medium/large systems, and architecture 3 because of its high cost (not coupled with significant performance improvements), the architectural choice is restricted to architectures 2 and 4. No crucial difference appears to exist between the two alternatives.

These considerations, together with the performance results presented in the previous section led to the choice of the architecture of TOMP. It was implemented following the structure of architecture 4 since it was felt that the performance improvement, particularly in the case of small system, would have paid the increase in complexity.

The next section will thus focus on the study of architecture 4, carried on by means of other modelling tools still based on the Markovian assumptions and workload model introduced at the beginning of this chapter.

2.2. OTHER MODELING TECHNIQUES AND MEASUREMENTS

2.2.1. Introduction

Architecture 4 was selected as the result of a case study in which a single modeling technique and a single set of assumptions were used to compare the performance characteristics of several processors-memories interconnection structures. A more detailed investigation of the chosen architecture was carried on using different modeling techniques each one best suited for a specific field of analysis. The new models are still based on Markovian assumptions and still assume the same workload described in Sections 2.1.2. and 2.1.3. Each modeling technique is best suited to a specific field or stage of the performance analysis.

Stochastic Petri Nets (SPN) are introduced first as a useful graphical tool for the precise description of the system operations, and as a modeling technique that allows a ready evaluation of the performance of small systems. SPNs are extremely versatile and are an excellent analysis tool in the initial stage of a study. They can be used by system designers with hardly any modeling experience and with no knowledge of the theory of Markovian processes since the model solution can be automatically obtained from the graphical description |MOLL81, NATK80|. Several design options can thus be tested for performance improvements with very little effort. A weak point of SPNs is that the graphical representation rapidly becomes very cumbersome when the size of the multiprocessor system to be modeled increases. Moreover the number of states of the Markovian process underlying the SPN representation of the system shows an exponential growth as a function of the number of system components.

Subsequent steps in the performance evaluation of a system can be done by using Queueing Network models (QN) for which a wide gamut of exact and approximate solution methods has been published in the literature |BRUE80, GRAH78|. Computationally efficient approximate solution methods exist that allow the analysis of fairly large systems, possibly with multiple bus interconnection structures. Models derived in this way have little versatility and modifications in the assumptions often lead to intractable queueing networks.

Simulation and measurements were employed to validate analytical results. Different from the previous approaches, simulation models can be very general and do not need the introduction of Markovian or independence assumptions. They can thus be used to investigate the impact that these assumptions have on the performance estimates provided by the other methods.

2.2.2. Stochastic Petri Net Models

Graph models of computer systems have been developed by many authors. Much of the work in this field refers to the original ideas of C. A. Petri and the derived models are generally known as Petri nets |PETE77, PETR66|. These models have recently become very popular for the representation of distributed computing systems because of their capability of clearly describing concurrency, conflicts, and synchronization of tasks.

Original Petri nets were used to represent the logic behaviour of computer systems with no timing and performance considerations. Some work has recently been done to include the notion of time into Petri net models. In some cases fixed times were used, leading to deterministic models |MERL76, ZUBE80|. Several authors have, on the other hand, suggested the introduction of random firing times in Petri nets |MOLL81, NATK80, SYMO80|, so that the resulting models are stochastic in nature.

In this section we refer to the work of M. K. Molloy |MOLL81|, and use the models he proposed, named Stochastic Petri Nets (SPN), to analyze the performance behaviour of some extensions of architecture 4.

A SPN has a set of places P, a set of transitions T, a set of directed arcs A, and an initial marking M. A firing rate in the set G is associated with each transition. A formal definition of a stochastic petri net is the following:

(2.21) $$SPN = (P,T,A,M,G)$$

where:

$$P = P_1 ,P_2 ,\ldots,P_n$$

$$T = t_1 ,t_2 ,\ldots,t_m$$

(2.22) $$A \subset (P \times T) \cup (T \times P)$$

$$M = m_1 ,m_2 ,\ldots,m_n$$

$$G = g_1 ,g_2 ,\ldots,g_m$$

In the graphical representation of stocastic Petri nets places are drawn as circles and transitions as bars. Arcs connect places to transitions and transitions to places. A marking is defined by assigning a given number m of tokens (drawn as black dots) to each place. A place is an input to a transition whenever an arc exists from that place to that transition. A place is an output of a transition whenever an arc exists from that transition to that place. A transition is enabled when all of its input places contain at least one token. Enabled transitions fire after an exponentially distributed random time. The rate associated to the transition

determines the parameter of the exponential distribution. The firing
of a transition removes one token from each of its input places and
puts one token in each of its output places. The firing of a
transition may thus disable conflicting transitions.

Molloy has shown that SPNs are isomorphic to continuous time
Markov Chain (MC); SPN markings correspond to MC states. It is
thus possible to obtain the steady state probability of each marking
and from it to derive performance indices.

Consider for instance the case of architecture 4 assuming that
the request for a common memory cannot preempt the local
processor. The requesting processor however can acquire the global
bus while waiting. A SPN model af the system in the case of two
processors is shown in Figure 2.17.

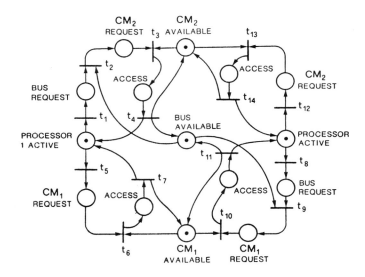

Fig. 2.17 – SPN of a two processor system in which the global bus
          is occupied also when the requested common memory is
          busy.

The assumptions on the multiple processor workload determine the
firing rate of each transition, giving:

(2.23)
$$g_1 = g_5 = g_8 = g_{12} = \lambda/2$$

$$g_4 = g_7 = g_{11} = g_{14} = \mu$$

The other transitions correspond to operations (bus arbitration and

release, acquisition of memory control, etc.) that, given the level
of detail considered in these models, are not relevant to the
modeling of bus contention and memory interference. We thus assume
these transitions to fire at a very high speed that we set equal to
α .   If more detailed models are necessary,   it is easy to
explicitly account for these phenomena by carefully setting the
corresponding transition speeds.

It can be shown that a MC model is derivable from the SPN of
Figure 2.17, whose solution provides the desired performance
measures.   The whole process can be made automatic and thus
transparent to the designer who only needs to define the SPN
topology and parameters.

Figure 2.17 is a SPN model of a system in which a processor
requesting to access a busy external common memory module can
nevertheless occupy the global bus. This may lead to a performance
loss when more than two processors are present in the system.   If,
instead, the bus arbiter prevents the processors from seizing the
bus unless the destination memory is free,   the SPN becomes a little
simpler: two places and two transitions are removed in the case of
a two processor system.   The resulting SPN is shown in Figure 2.18
in which the firing rates of the remaining transitions are left
unchanged with respect to the previous case.

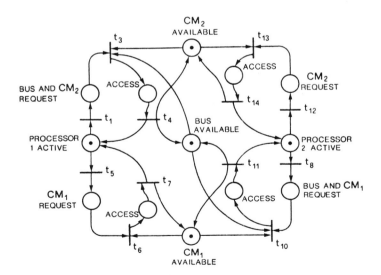

Fig. 2.18 – SPN a two processor system in which the global bus is
        occupied only when the requested common memory is
        available.

Three things should be made clear from these two examples. First of all SPNs provide a way of obtaining a simple and precise description of the system operation and workload. Secondly, an automated Markovian analysis of the system is possible without having to go through the process of finding the correct state definition and of evaluating the state transition rates. Finally SPNs are extremely versatile: a change in the assumptions is often immediately represented with a little modification of the net.

To further emphasize these concepts, let us consider a more detailed view of the system operations. Assume the common memory accesses to be actually subdivided into a set of elementary memory operations so that some degree of interleaving is possible: processors trying to access the same memory module will be served in parallel at a rate smaller than μ, but larger than $\mu/2$. The SPN that describes this behaviour is easily drawn in the case of a two processor system and is shown in Figure 2.19. Since the lengths of the two accesses are independent, identically distributed random variables, the probability of either access ending first is 0.5. We thus have two conflicting transitions, t_9 and t_{10} in one case,

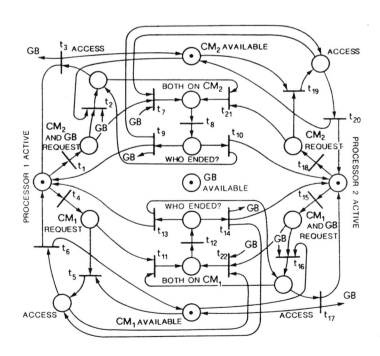

Fig. 2.19 – SPN of a two processor system with interleaving of memory accesses.

and t_{13} and t_{14} in the other case, that select which processor must continue accessing the memory while the other returns to its active state. The rates of these transitions are equal, so that they fire with equal probability. Transition rates are in this case assigned as follows:

(2.24)

$$g_1 = g_4 = g_{15} = g_{18} = \lambda/2$$

$$g_3 = g_6 = g_{17} = g_{20} = \mu$$

$$g_8 = g_{12} = \mu'$$

All other transition rates are set to the already mentioned very high speed α to make negligible the times needed for those operations to occur in our models. The rate μ' represents the speed at which two simultaneous access requests to the same common memory module are served. This rate is contained in the range $(\mu, 2\mu)$ since no interleaving keeps the speed equal to the rate of service of a single request and full interleaving can make the rate twice as large.

A range of operational conditions can be easily inspected exploiting the versatility of the SPN model. For instance, if we eliminate transitions t_{10} and t_{13} and set $\mu' = \mu$ the SPN reduces to the case in which preemption of the access performed by a processor on its local memory is allowed (i.e. requests coming from external processors through the global bus are given priority).

If, instead, in the SPN of Figure 2.19, we set $\mu' = 2\mu$ the case of dual port memories that can support simultaneous accesses is represented. We thus obtain the SPN model of architecture 3.

We have already said that SPNs are a powerful tool for a precise representation of distributed systems characterized by parallelism, contention, and queueing. As any other modeling tool, however, they also have weak points, such as the practical difficulty of representing systems of increasing size, and the rapid growth of the number of states of their isomorphic Markov chains.

As an example consider the extension of the SPN of Figure 2.17 to a three processor case which yields the graph of Figure 2.20. The SPN becomes rather complex despite the introduction of the simplifying assumption of managing the global bus queue with a random selection policy. From a conceptual point of view the SPN representation of the three processor system is no more difficult than that of the two processor case. In fact, a modular description of the SPN can be given. The replication of modules with identical structure must be provided, together with a precise description

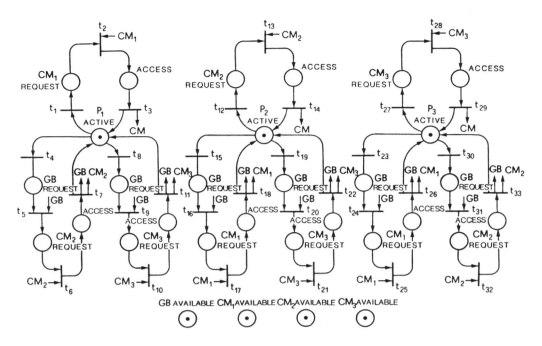

Fig. 2.20 – SPN of a three processor system as in Figure 6.

of the interconnection among them. The increased complexity of the graphical representation of the three processor system is followed by an explosion of the number of states of the associated Markov chain. From 26 states in the 2 processor case we go to 395 states in the 3 processor case. The Markov chain is still solvable, thus the analysis of the system is still feasible, but this growth indicates that even slightly larger systems become intractable.

The processing power P of a two processor system is presented in Table 2.V. Different values of the load factor are considered as listed in the first column. The second column contains the processing power values obtained from the SPNs of Figures 2.17, 2.18, and 2.21. The SPNs of Figures 2.17 and 2.18 lead to different processing power values only for systems comprising more than two processors; the SPN of Figure 2.19 reduces to that of Figure 2.17 when no interleaving is considered. The third column contains the processing power values provided by the SPN of Figure 2.19 when full interleaving is assumed by setting $\mu' = 2\mu$. The last column contains the processing power values obtained with the assumption that external processors may preempt local accesses on common memories.

Table 2.V
Processing power of a two processor system obtained from (a) the
SPN's of Figures 2.17–2.19, (b) the SPN of Figure 2.19 with $\mu'=2\mu$,
and (c) considering preemption.

ϱ	a	b	c
0.01	1.980	1.980	1.980
0.05	1.901	1.904	1.901
0.1	1.807	1.814	1.807
0.2	1.633	1.655	1.633
0.4	1.346	1.400	1.350
0.6	1.131	1.208	1.139
0.8	0.968	1.059	0.979
1.0	0.842	0.941	0.857

These results show how changes in the assumptions have a limited
impact on processing power values of a two processor system,
particularly when the system load is low. More significant
differences are expected for larger system.

2.2.3. Queueing Network Models

An alternate approach for the analysis of extensions of the
architecture 4 is an explicit representation of the contention for
shared resources with a queueing network (QN). Figure 2.21 depicts
a QN model of the multiprocessor system in which queueing may
occur at the global bus and at the common memory modules.
Processors are represented as delay elements. Common memory access
requests can be viewed as customers that require service from the
common memory modules. Upon completion of their service, access
requests return to their corresponding processor stations where they
are delayed before repeating the memory access cycle. This delay
represents processor activity between subsequent access requests
(CPU burst).
        One class of QN models for which computing the exact solution
is simple is that of the so called product form QNs considered by
the BCMP theorem |BASK75|. Simultaneous possession of resources
(for which individual queues exist) and priority queueing

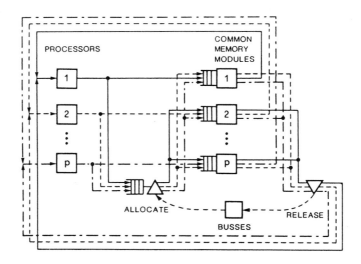

Fig. 2.21 – Queueing network representation of a generalisation of
              the TOMP architecture with a multiple bus intercon-
              nection network.

policies normally make QN models of multiprocessor systems not
product form. Nice computational properties of approximate solutions
of non product form QN models can however be exploited when some
additional restrictions are imposed on the behaviour of the system
represented in Figure 2.21. Let us assume that processors trying to
reach an external common memory module, having found the global
bus free, seize it and, if necessary, wait in queue for the
destination memory module to become available without releasing the
global bus. Under these operating conditions the QN of Figure 2.21
becomes a model with a passive resource (the global bus) |CHAN78,
KELL76|. Approximate solutions of these models can be obtained by
using decomposition and equivalence techniques |BALB82, CHAN75a,
CHAN75b| that are standard in the QN field.
       Let us identify the common memory modules as a single common
memory subsystem loaded by the processors with memory access
requests. At any point in time up to p access requests can be
queued or receiving service from the memory subsystem, but only
one of them can be an external request (only one bus is
available). An aggregated model such as that of Figure 2.22 can be
obtained replacing the memory subsystem with an equivalent station
whose service characteristics are computed using a set of controlled
experiments |CHAN78, DENN78|.
       A further simplification can be obtained when the identity of
the access requests (issuing processor identifier) is removed and a

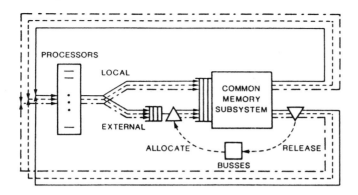

Fig. 2.22 – Flow equivalent representation of the memory subsystem.

set of requests belonging to two different classes (the class of local requests, and the class of external requests) is assumed to load the common memory subsystem.   This abstract view of the multiprocessor system can be used to represent more general cases in which the architecture   here   considered   is   extended   to   embody   an   inter-connection network comprising an arbitrary,   but finite number of b busses. The resulting model is that of Figure 2.23.

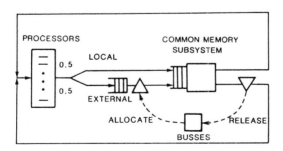

Fig. 2.23 – Aggregate   memory   representation   of   the   queueing
          network of Figure 2.21.

The success of this aggregation technique relies on the possibility of replacing the common memory subsystem with a memory flow equivalent station so that the behaviour of the processors is left unchanged by such a substitution.   The functional characteristics of the equivalent service station are obtained by computing the response time of the memory subsystem to access requests of the two

classes, assuming the memory subsystem itself to be kept under a
fixed load. This technique is known to be exact when the original
model is of the product form type |BALB82, CHAN75a, VANT78|, while
provides approximate results in the other cases |CHAN75b, SAUE81|.
The controlled experiments for the computation of the parameters of
the equivalent memory station are performed on a model that can be
obtained from the original QN (Figure 2.21) by shortcircuiting the
processor and bus servers, and by assuming that fixed mixes of
memory access requests are continuously asking for service from the
memory modules. Figure 2.24 depicts the multiclass network used to
perform the controlled experiments in the general case of a p
processor system. Each class in this model is a closed path
throughout the network used by a fixed number of customers. A
class is said to be alive if at least one customer of that class is
present in the model and it is said to be dead otherwise. In the
case of the model of Figure 2.24, alive classes have a population of
only one customer. The first p classes represent local common
memory requests. A fictitious station (with zero mean service time)
is added to the model to accumulate the global throughput of local
requests provided by the memory subsystem.

Local requests do not interfere among each other since they

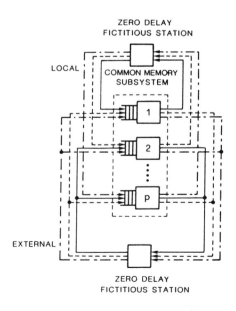

Fig. 2.24 – Multiclass model for the computation of the parameters
       of the equivalent station.

visit different memory modules. The second set of p classes is introduced to represent the external common memory requests. Each one of these classes visits p-1 memory modules. The global behaviour of the external access requests is collected adding another fictitious station with zero mean service time (see Figure 2.24).

A request mix is characterized by having n classes (n=1,2,...,p) simultaneously alive. The symmetry of the system allows to identify a request mix with any distribution of n requests over the 2p different classes with the constraints that classes i and p+i (i=1,2,...,p) cannot be simultaneously alive (a processor issues either a local request or an external request), and that the number of external classes simultaneously alive cannot exceed the number of available busses b. Let n_l and n_e represent the number of local and external classes that are alive in a given request mix. A request mix is identified by the pair (n_l,n_e). Two request mixes are considered different when the corresponding pairs are different.

Let $X_l(n_l,n_e)$ represent the throughput of the first fictitious station when the memory subsystem is loaded with the (n_l,n_e) request mix; similarly let $X_e(n_l,n_e)$ represent the throughput of the second fictitious station under the same subsystem load. These two quantities represent the rates of local and external requests satisfied by the memory subsystem under given loads. Their reciprocals can be interpreted as mean response times. The behaviour of the memory subsystem (the service function of the equivalent memory station) is represented by the set of mean response times obtained by loading the memory subsystem with all the possible request mixes.

The computation of the service function of the memory equivalent station is completed assuming that the memory subsystem responds with a rate of service $X(n_l,b)$ every time it is actually loaded with a request mix (n_l,n_e) such that n_e b. When the bus interconnection network is made up of p busses, and thus behaves as a crossbar switch, the original model is a product form network and the aggregation process is exact. When the number of busses (b) is smaller than the number of processors (p) the aggregation process leads to an approximate QN model of the multiprocessor system that in general does not satisfy the conditions for a product form solution |BALB82|, but that can nevertheless be solved with an efficient technique proposed by Herzog |HERZ75|. The accuracy of the approximation degenerates with the decreasing of the number of busses and has a worst case represented by the single global bus interconnection structure.

The results of this approximate analysis of the multiple bus model shown in Figure 2.23 have been successfully validated with a

Table 2.VI
Processing power of a three processor system in the case of 1,2, and 3 buses.
Comparison of the approximate analytical results obtained with a queueing network
against the interval estimates obtained with simulation. In the case of 1 bus, the
exact solution obtained with the SPN of Figure 2.6 is also presented.

ϱ	1 BUS				2 BUSSES			3 BUSSES		
	SPN	QN	simulation low	up	QN	simulation low	up	QN	simulation low	up
0.05	2.850	2.850	2.70	2.99	2.853	2.71	2.99	2.853	2.70	2.99
0.1	2.704	2.701	2.56	2.83	2.713	2.58	2.85	2.713	2.57	2.84
0.2	2.427	2.419	2.29	2.53	2.455	2.34	2.58	2.456	2.38	2.58
0.3	2.189	2.167	2.05	2.26	2.228	2.13	2.35	2.230	2.12	2.34
0.4	1.963	1.948	1.85	2.04	2.030	1.94	2.14	2.033	1.93	2.13
0.5	1.776	1.759	1.73	1.91	1.859	1.78	1.96	1.862	1.78	1.96
0.6	1.616	1.598	1.57	1.73	1.709	1.61	1.77	1.714	1.63	1.80
0.8	1.358	1.340	1.28	1.41	1.466	1.38	1.52	1.473	1.39	1.53
1.0	1.163	1.147	1.11	1.22	1.278	1.21	1.33	1.286	1.21	1.33

Table 2.VII
Processing power of a four processor system in the case of 1,2, and 3 buses. Comparison of the approximate analytical results obtained with a queueing network against the interval estimates obtained with simulation.

ϱ	1 BUS			2 BUSSES			3 BUSSES			4 BUSSES		
	QN	simulation low	up	QN	simulation low	up	QN	simulation low	up	QN	simulation low	up
0.05	3.797	3.61	3.99	3.803	3.62	3.99	3.803	3.62	3.99	3.803	3.62	3.99
0.1	3.592	3.42	3.78	3.615	3.44	3.80	3.615	3.43	3.79	3.615	3.44	3.80
0.2	3.199	3.03	3.34	3.266	3.10	3.41	3.269	3.13	3.46	3.269	3.13	3.46
0.3	2.846	2.71	2.99	2.957	2.80	3.09	2.963	2.82	3.11	2.963	2.79	3.08
0.4	2.542	2.39	2.64	2.686	2.45	2.70	2.698	2.58	2.84	2.698	2.58	2.84
0.5	2.283	2.14	2.36	2.451	2.34	2.59	2.467	2.40	2.66	2.467	2.38	2.61
0.6	2.065	1.97	2.17	2.247	2.10	2.31	2.267	2.16	2.38	2.267	2.15	2.37
0.8	1.723	1.62	1.79	1.915	1.82	2.01	1.941	1.82	2.01	1.943	1.90	2.10
1.0	1.472	1.36	1.50	1.660	1.60	1.77	1.691	1.64	1.81	1.692	1.64	1.81

detailed simulation of the original system model. Tables 2.VI and
2.VII present the validation results in which simulation interval
estimates are obtained using the regenerative method |IGLE78,
LAVE78| with 95% confidence interval and 5% width. The
approximation of the analytical results is, in the cases we have
considered, extremely good so that they are covered by the
confidence intervals provided by the simulation. In the analysis of
a three processor system and single global bus, the SPN model of
Figure 2.20 provides exact results that can be used to check the
behaviour of the approximation method in worst case conditions. The
very good agreement found between the results obtained with the
two techniques suggests the possibility of using this fast
aggregation method in the analysis of larger size systems.

2.2.4. Measurements

Measurements have been performed |AJMO83| on the three processor
TOMP prototype shown in Figure 2.25. Each processor runs a cyclic

Fig.2.25 – View the present version of the TOMP prototype.

process corresponding to the sequential execution of a CPU burst
and a common memory access. The CPU burst consists of a sequence
of operations on the private memory; let T_p represent the time
required for the execution of each elementary operation, and N_p the
number of operations in a sequence. A common memory access is a

sequence of elementary read/write operations on a common memory area that can be either local or external; each elementary operation requires a time T_a' when directed to the local common memory module and a time T_a'' when directed to an external common memory module. The number of elementary operations needed to perform an external access sequence is N_a. Following the assumptions previously introduced, the probabilities of accessing local or external common memory modules have been respectively set to the following values:

$$\text{Pr } |\text{accessing local common memory}| = 0.5$$

(2.25)

$$\text{Pr } |\text{accessing each external common memory}| = 0.25$$

The aim of the measurements was to validate the stochastic models used in the performance analysis of the system, with respect to the study of contention for physical resources rather than to the modeling of applications or system control software. For this reason the workload applied to the real system closely approximates the model assumed in the previous section. The quantities Np and Na approximate exponentially distributed random variables with average values $E|N_p|$ and $E|N_a|$, respectively.

The experimental values of processing power were obtained with direct measurements of three signals on each processor module. These signals (named GREQ, GSERV and LREQ) are active whenever the following conditions hold:

GREQ : the processor has requested the global bus, but access has not been granted yet.

GSERV: the processor is using the global bus to access an external common memory module

LREQ : the processor is either waiting to access its common memory module or accessing it

The measurement instrumentation (see Figure 2.26) is composed of a 5 MHz pulse generator, a simple gating logic circuit and a set of pulse counters. Defining f_m to be the pulse frequency, and N_x to be the number of pulses measured by the counter associated to signal x over a fixed measure time T_m, we obtain the fraction of time during which signal x is active, $F(x)$, as:

(2.26)
$$F(x) = \frac{N_x}{f_m T_m}$$

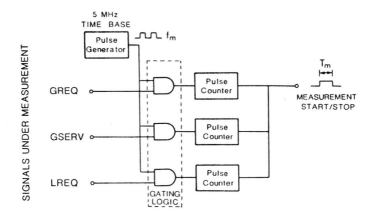

Fig. 2.26 - Block diagram of the measurement setup.

The fraction of time during which processor i is not active, $F_i(NACT)$, can be evaluated as:

(2.27) $F_i(NACT) = F_i(GSERV) + F_i(GREQ) + F_i(LREQ)$

and the processing power is obtained as:

(2.28) $P = 3 - F_i(NACT)$

For each measurement the value of the load factor ϱ may be estimated as:

(2.29) $\varrho = \dfrac{N_a T_a}{N_p T_p}$

where T_a is the common memory access time. Since in the TOMP prototype the access time to a common memory module has different values for the local and external processors we assume (as a first order approximation) for T_a the mean of the two values T_a' and T_a'', thus:

(2.30) $T_a = \dfrac{T_a' + T_a''}{2}$

The measured values of processing power, obtained as in

(2.28),versus   the   load   factor   as   given   by   equation   (2.29)   are
plotted in Figure 2.27, curve (a).

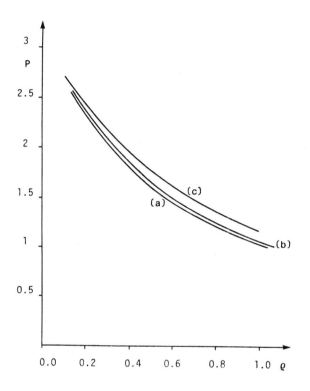

Fig. 2.27 – Processing   power   of   a   three   processor   system.
            Comparison of the analytical results obtained with the
            SPN of Figure 2.20 |curve (c)| against the measurement
            results |curves (a) and (b)|.

Several approximations introduced in the description of the system
workload and some architectural features that were not taken into
account in the models, make the previously defined parameter
not exactly represent the real load of the multiprocessor system.
The impact of these approximations on the estimate of the parameter
can be substantial, and a more careful derivation of its value is
needed. In the following paragraphs we discuss some of these
phenomena to derive a new formula for the evaluation of the load
factor.

A - Non exponential distributions
      At the beginning of each CPU burst each processor computes

three uniformly distributed random numbers by means of a shift
register and of an irreducible polynomial of order 32 |PETE61|.
The first random number is 2 bit wide and it is used to
determine which common memory area is to be accessed during
the next access period. The other two random numbers are used
as pointers into logarithmic conversion tables to obtain the two
random numbers N_a and N_p. Therefore the exponential
distributions of the CPU burst and of the access period are in
fact approximated by truncated geometric distributions. N_a can
vary between 1 and 2^{16}, and N_p can vary between a minimum
number N_c and a maximum $N_c + 2^{16}$. N_c is the number of
internal cycles needed to evaluate the three random numbers.

B – Internal operations

Individual common memory accesses, although being part of a
single sequence, are separated by a short gap of duration T_i
corresponding to processor operations (register transfers); this
fact allows a partial interleaving of the operations of different
processors on the global bus and increases the mean CPU burst
duration by a factor proportional to the mean access time. We
can then correct the load factor ϱ as follows:

$$(2.31) \qquad \varrho\,' = \frac{N_a\,T_a}{(N_p\,T_p + N_a\,T_i)}$$

C – Memory refresh

The private memory of the prototype processors was implemented
using dynamic memories refreshed by the processor itself. The
refresh operations are carried on in parallel with the execution
of the workload process, and their effect is to increase the CPU
burst duration by a factor T_r/T_{ir}, where T_r is the time
required to execute a refresh operation and T_{ir} is the time
interval between two successive refresh operations. Therefore
the load factor can be further corrected as:

$$(2.32) \qquad \varrho\,'' = \frac{N_a\,T_a}{N_p\,T_p\,(1 + T_r/T_{ir}) + N_a\,T_i}$$

D – Global bus arbitration

The arbitration is carried on by a distributed synchronous
logic that assigns the bus to one processor at a time following

a fixed priority policy. The arbitration process lasts a certain amount of time during which all the requesting processors are held waiting. This time can vary between two limits T_{armn} and T_{armx}. In order to consider the global bus arbitration on the load factor ϱ we can add to each external common memory access time T_a", the average bus arbitration time, T_{ar}, extimated as:

$$(2.33) \qquad T_{ar} = \frac{T_{armn} + T_{armx}}{2}$$

hence we obtain that the corrected common memory access time T_a" is given by:

$$(2.34) \qquad T_a''' = \frac{T_a' + T_a'' + T_{ar}}{2}$$

and the load factor ϱ can be finally corrected obtaining

$$(2.35) \qquad \varrho''' = \frac{N_a\ T_a'''}{N_p\ T_p\ (1 + T_r/T_{ir}) + N_a\ T_i}$$

The measured values of processing power versus the corrected load factor ϱ''' are shown in Figure 2.27, curve (b). Curve (c) in the same figure presents the results given by the SPN of Figure 2.20. The comparison between the measured values and the analytical model estimates confirms the validity of the modeling approach we have used. The performance predictions are very accurate in spite of the differences between the actual prototype behaviour and the assumptions on system operations used in the development of the model, where an abstract view of the system behaviour neglects the actual details of the elementary operations executed by the prototype.

2.3. REFERENCES

|AJMO81| Ajmone Marsan,   M.;   and Gregoretti,   F.   "Memory
              Interference Models for a Multiprocessor Systems,   with a
              Shared Bys and a Single External Common Memory",
              EUROMICRO JOURNAL, February 1981.

|AJMO82a| Ajmone Marsan,   M.,   and Gerla, M. "Markov   Models for
              multiple Bus Multiprocessor Systems",   IEEE Trans.   on
              Computers, March 1982.

|AJMO82b| Ajmone Marsan,   M.;   Balbo,   G.;   and Conte,   G.
              "Comparative   Performance   Analysis   of   Single   Bus
              Multiprocessor Architectures", IEEE Trans. on Computers,
              December 1982.

|AJMO83| Ajmone Marsan,   M.;   Balbo,   G.;   Conte,   G.   ;   and
              Gregoretti,   F.   "Modeling Bus Contention and Memory
              Interference in a Multiprocessor System", IEEE Trans. on
              Computers, January 1983.

|BALB82| Balbo,   G.;   and Bruell, S.C., "Computational Aspects of
              Aggregation   in   Multiple   class   Queueing   Networks",
              Performance Evaluation, August 1983.

|BASK75| Baskett, F.; Chandy, K.,M.; Muntz, R.R.; and Palacios,
              F.,G. "Open, Closed and mixed networks of queues with
              different   class   of   costumers",   Communication   Ass.
              Computing Mach., June 1976.

|BASK76| Baskett,   F.;   and Smith,   A.J.   "Interference   in
              Multiprocessor   Computer   Systems   with   Interleaved
              Memory", Communications of the ACM, June 1976.

|BHAN75| Bhandarkar,   D.P.   "Analysis of Memory Interference in
              Multiprocessors",   IEEE   Transactions   on   Computers,
              September 1975.

|BRUE80| Bruell,   S.,   C.;   and Balbo, G. "Computational Algorithms
              for Closed Queueing Networks". Amsterdam,
              The Netherlands; Elsevier, 1980.

|BUZE77| Buzen, J.P.; and Potier, D. "Accuracy of the Exponential
              Assumption in Closed Queueing Models", Proceedings 1977

SIGMETRICS/CMG International Conference on Computer Performance, Modeling, Measurement, and Management, Whashington D.C., November 1977.

|CHAN75a| Chandy, K.M.; Herzog, U.; and Woo, L. "Parametric Analysis of Queueing Network Models", IBM J. of Res. and Dev., January 1975.

|CHAN75b| Chandy, K.M.; Herzog, U.; and Woo, L. "Approximate Analysis of General Queueing Networks", IBM J. of Res. and Dev., January 1975.

|CHAN78| Chandy, K.M.; and Sauer, C.H. "Approximate Methods for Analyzing Queueing Network Models of Computer Systems", ACM Computing Surveys, September 1978.

|CHAN80| Chang, S.S.L. "Multiple-Read Multiple-Write Memory and Its Applications", IEEE Trans. on Computers, August 1980.

|CONT81| Conte, G.; Del Corso, D.; Gregoretti, F. and Pasero, E. "TOMP80: a Multiprocessor Prototype", EUROMICRO 81, Paris, September 1981.

|DENN78| Denning, P.J.; and Buzen, J.P. "The Operational Analysis of Queueing Network Models", ACM Computing Surveys, September 1978.

|FUNG79| Fung, F.; and Torng, H. "On the Analysis of Memory Conflicts and Bus Contentions in a Multiple-Microprocessor System", IEEE Trans. on Computers, January 1979.

|GRAM78| Graham, G., S. Guest Editor "Special Issue: Queueing network models of computer systems" Ass. Computing Mach. Comp. Survey, September 1978.

|HERZ75| Herzog, U.; Woo, L.; and Chandy, K.M. "Solution of Queueing Problems by a Recursive Technique", IBM J. of Res. and Dev., May 1975.

|HOEN77| Hoener, S.; and Roeder, W. "Efficiency of a Multiprocessor System with Time-Shared Busses", EUROMICRO 77, September 1977.

|HOOG77| Hoogendoorn,   C.H.   "A   General   Model   for   Memory
              Interference   in   Multiprocessors",   IEEE   Trans.   on
              Computers, October 1977.

|IGLE78| Iglehart,   L.D,   "The regenarative Method for Simulation
              Analysis" in Current Trends in Programming Methodology,
              Vol.iii,   K.M.   Chandy;   R.T.   Yeh Eds.,   Prentice Hall,
              1978.

|KAIS80| Kaiser,   D.   "iAPX 432 Object Prime Preliminary Draft",
              Intel Corporation, August 1980.

|KELLE6| Keller,   T.W.   "Computer   System   Models   with   Passive
              Resources",   Ph.D.   Thesis, Computer Science Department,
              University of Texas at Austin, 1976.

|KEME60| Kemeni,   J.G.;   and Snell,   J.L. "Finite Markov Chains",
              Van Nostrand, Princeton, 1960.

|KLEI75| Kleinrock,   L.   "Queueing Systems.   Vol.   1",   J. Wiley,
              New Jork, 1975.

|LAVE78| Lavenberg,   S.S.   "Regenerative Simulation of Queueing
              Networks" IBM Research Report RC-7087, 1978.

|LEVY78| Levy,   J.V. "Buses, The Skeleton of Computer Structures",
              Computer Engineering: a DEC View of Hardware System
              Design by C.G.   Bell,   J.C.   Mudge, and J.E. McNamara,
              1978.

|MERL76| Merlin,   J.A.;   and Farber,   D.J.   "Recoverability of
              Communication Protocols - Implications of a Theoretical
              Study" IEEE Trans. on Commun., September 1976.

|MOLL81| Molloy,   M.K.   "On   the   Integration   of   Delay   and
              Throughput Measures in Distributed Processing Models",
              PhD Thesis, University of California, Los Angeles,   1981.

|NATK80| Natkin,   S.   "Reseaux de Petri Stochastiques" These de
              Docteur-Ingegneur, CNAM-PARIS, June 1980.

|PALM58| Palm,   c.;   "The Assignement of Workers in Servicing
              Machines",   The   Journal   of   Industrial   Engineering,
              September 1958.

|PATE79| Patel, J.H. "Processor–Memory Interconnections for Multiprocessors" Proc. 6-th Annual Symposium on Computer Architecture, April 1979.

|PETE61| Peterson, W.; and Weldon, E. "Error Correcting Codes", MIT Press, 1961.

|PETE77| Peterson, J.L. "Petri Net Theory and Modeling of System", Prentice Hall Inc. Englewood Cliff N.J., 1981.

|PETR66| Petri, C.A. "Communication with Automata", PhD Thesis, Technical Report RADC–TR–65–377, New York, January 1966.

|SAUE81| Sauer, C.H "Approximate Solution of Queueing Networks with Simultaneous Resource Possession", IBM Research Report RC 8679, January 1981.

|SETH77| Sethi, A.S.; and Deo, N. "Interference in Multiprocessor Systems with Localized Memory Access Probabilities", IEEE Transactions on Computers, February 1979.

|SYMO80| Symons, F.J.W. "Introduction to Numerical Petri Nets, a General Graphical Model of Concurrent Processing Systems", A.T.R., January 1980.

|SWAN77| Swan, R.J.; Fuller, S.H.; and Seviorek, D.P. "CM*; a Modular Multimicroprocessor", Proc. AFIPS National Computer Conference, 1977.

|THUR72| Thurber, K.J.; Jensen, E.D.; Jack, L.A.; Kinney, L.L.; Patton, P.C.; and Anderson, L.C. "A Systematic Approach to the Design of Digital Bussing Structures", Proc. AFIPS Fall Joint Computer Conference 41, 1972.

|VANT78| Vantilborgh, H. "Exact Aggregation in Exponential Queueing Networks", Journal of the ACM, October 1978.

|WILL78| Willis, P.J. "Derivation and Comparison of Multiprocessor Contention Measures", IEE Journal of Computers and Digital Techniques, August 1978.

|WULF72| Wulf,   W.A.;   and   Bell,   G.C.   "C.mmp,   a
Multiminiprocessor",   Proc.   AFIPS Fall Joint Computer
Conference, 1972.

|ZUBE80| Zuberek,   W.M.   "Timed Petri Nets and Preliminary
Performance Evaluation",   Proc.   of the 7-th Annual
Symposium on Computer Architecture, 1980.

CHAPTER 3

TOMP SOFTWARE

F.Gregoretti
Dipartimento di Elettronica
Politecnico di Torino
Torino, ITALY

ABSTRACT. This chapter describes TOS, the Operating System of the
TOMP prototype. First it outlines the motivations, the general
structure, the interprocess communication model and primitive
operations. Then it describes in more detail the functions of the
ROM resident kernel, the peripheral handling strategy and it
presents the architecture of the debugging environment. Finally a
critical review of the software project is given.

3.1. INTRODUCTION

The structure of the system software for any computer, but even
more for a multiprocessor depends both on the architecture of the
machine and on the computational model the designer intends to
support. The choice and the definition of such a model is complex
because the term parallel processing or parallel software covers a
very broad spectrum of topics, ranging from designing algorithms
for systolic arrays to communications on local networks.
Nevertheless it contains the general basic idea of "programming" a
set of processing elements to cooperate in parallel to the execution
of a common task. The goals of parallel processing may be the
execution speed or reliability or the fault tolerance. The success in
achieving them depends on the machine architecture, the operating
system, the application, its partitioning and over all on the
reciprocal matching of all these components. Jones and Schwarz
|JONE80| have analyzed the spectrum of parallel applications and
have characterized it according to three parameters, in order to
establish the fitness of a multiprocessor machine to a parallel
solution.

G. Conte and D. Del Corso (eds.), Multi-Microprocessor Systems for Real-Time Applications, 87–116.
© 1985 by D. Reidel Publishing Company.

- <u>Computational Unit</u>. It is the minimum sequence of operations that is performed in parallel with other ones. Computation units may be homogeneous or heterogeneous. Multiprocessors are composed of independent processors, each capable of executing a different program. Therefore for them the execution unit is the process.

- <u>Communication patterns</u> which establish the requirements of the computation units for exchanging control information (synchronization) or data. Of particular interest are the frequency of the communication and the amount of data which are exchanged. Because of the shared memory area, multiprocessors are well suited for tranferring bulk data between processes. On the other hand for multiprocessors the frequency of interaction is critical because of the high cost of creating, suspending, resuming the process unit.

- <u>References to data</u>. As they are well suited for non-homogeneous processes, multiprocessors fit also well patterns of reference to data which are irregular, non deterministic or dependent on the status of the data or of the computation.

Therefore the computational paradigm which best matches the architecture of a multiprocessor machine and is supported by many parallel operating systems |DOWS79, JONE79, OUST80| is that of a set of processes, possibly etherogenous, each one allocated to a different processor. They cooperate towards a common goal with a degree of parallelism far larger than the single instructions of the processors and may communicate in a <u>controlled</u> way large and non a-priori predictable amounts of data. Such a model, matches most of the applications for multiprocessors |JONE80a| but is nevertheless still too general in order to define in a precise way the characteristics of the software support that it requires. Applications for multiprocessors range from solving computationally hard problems, like PDE or molecular dynamics, to robotics and machine automation. The first ones are characterized by a large number of relatively small processes which are created and deleted dynamically and the main goal is to reduce the total execution time. In the latter ones the number of processes is small and they are relatively static within the life of the the application and often the goal is not to reduce the overall execution time, but to match the real time response to external stimuli.

Before making a choice we did not have a particular application in mind, but we considered factors like the size of the

prototype machine (< 10 processors),   the requirements of the
industrial environment towards which the project was directed and
the constraints which will be outlined in section 3.1.2.   Then we
decided to provide support for a model with a limited number of
processes,   statically allocated to processors, with a large grain of
parallelism and the requirement for a fast and efficient exchange of
information.
        In the following paragraphs of this section we will outline the
goals of the software project for TOMP-80,   the limitations in the
design and their reasons and we will give a schematic outline of
the environment.   In Section 3.2 we will describe the communication
mechanism.   Section 3.3 will describe the operations of the resident
executive and Section 3.4 the debugging and monitoring support.   In
Section 3.5 we will analyze utility processes and in Section 3.6 the
system generation facilities.   Finally in Section 3.7 we will make a
critical   analysis   of   the   whole   design   and   of   the   future
developments.

3.1.1. Goals and Motivations

Although the TOMP project has been mainly an architectural and
hardware oriented,   a limited software project was started in
parallel, having a number of goals.
        During the development of a new multiprocessor system there
is the necessity for software aids which can be made available
rapidly,   are tailored to the designer requirements,   and require a
limited effort and budget.   They have to be sufficiently flexible to
support major modification of the underlying hardware structure
during the system development.   The first goal of the software
project was therefore to give the basic support for the functional
verification of hardware modules.
        A second goal was to provide the system with a modular run
time support for the experimentation with parallel processes.   This
made it possible to construct a set of real or synthetic parallel
workloads for evaluating the performances of the architecture under
different load conditions.   It was in particular interesting to
validate with experimental results the predictions obtained by
analytic modelling tools |AJMO83|.
        The third goal was to provide a modular testbed for the
experimentation and evaluation of interprocess communication
patterns of different complexity,   ranging from the simple use of the
common memory,   to interprocessor interrupts and to the
implementation in hardware of part of the communication primitives.

The final goal was to investigate the methodologies for debugging parallel processes on a multiprocessor and to integrate the debug aids as a basic component of the software support.

3.1.2. Limits

In summary the software project for the TOMP-80 has been more conceived as a tool for the design and refinement of the architecture and as a research workbench than as a complete and engineered multiprocessor operating system for application development. This has been due also to two severe constraints for the project. First the implementation had to follow and partly support the hardware development; the resources devoted to this task were about one person at full time. Secondly only limited software development tools were available. In fact the first version of the software was developed using a cross-development package running on a 8-bit microcomputer system (MCZ 1/20), where assembly language only was available. More recently the PLZ-SYS high level language was made available for the 16-bit processors used in the prototype and only the latest modules use it.

Therefore it was necessary to narrow the scope of the research in the implementation phase and to introduce several simplifications in the design in order to meet the prototype development deadline (9 months). Although the design which will be presented here is nearly completed not all the facilities have been implemented with their full capabilities.

Nevertheless we think that, besides the use we have made during the prototype development, the software project has been of great utility in a better understanding of the potential performances and bottlenecks of the TOMP architecture and the interactions and tradeoffs between hardware and software.

3.1.3. Overall System Description

The software environment provided for the TOMP prototype is composed of an integrated set of tools, in the following referred as **TOS** (Tomp Operating System), whose main characteristics are:

- Modularity, with the possibility of a separate and independent updating of each system component. The actual multiprocessor is a prototype system and it is likely to be frequently updated in the near future. Hence hardware and software

flexibility to different architectures have been major design
issues. Moreover a modular design allows a smooth expansion
of the system to meet increasing requirements with minor
changes both in hardware and in software.

- Transparency of the multiprocessor architecture at the
  application level. This means that the programmers do not
  need to know on which processor an application process will
  be run and communication between processes take place on the
  basis of a global naming scheme, independent of the physical
  allocation of processes to processors. This allocation is static
  and determined off-line by a System Generation program
  running on a host computer; there is no possibility of
  dynamic creation or deletion of processes at run time.

- Integration of low-level monitoring and debug functions in the
  executive. These functions support higher level debugging
  tools running as system processes in parallel with application
  programs.

Although TOS is a completely integrated structure it can be
analyzed by dividing it into five major components:

- The **ICS** (Interprocess Communication System) which handles all
  the communication between system and application processes.
  It is composed by a set of procedures replicated in each
  processor and by common data structures;

- The **Executive,** comprising of several identical modules
  (kernels) each one residing on a processor. Its main
  responsibilities are process scheduling, interrupt and
  exception handling and providing low level debugging
  primitives;

- The **Debugger** which is composed of several low level modules
  (Local Monitors) interacting with the executive and with an
  upper level, the Debugger Interface providing the user with
  an interactive, high level interface. All these modules are run
  as system processes and communicate by means of the ICS;

- The **Utilities** which provide the main facilities for interfacing
  with system peripherals;

- The **System Generation Program** which runs on a host computer and is intended to support the user in the task of allocating logical objects, like processes or communication data structures, to physical resources, like processors or shared memory areas.

A pictorial view of TOS main components is given in Figure 3-1.

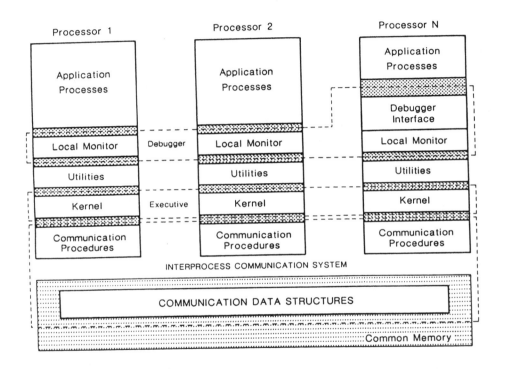

Fig. 3-1: Pictorial view of TOS main components.

- Distributed communication procedures and shared data for communication allocated to common memory.
- Executive composed of the local kernels of the processors.
- Debugger composed of the Local Monitors and the Debugger Interface.
- Utility processes distributed over the system.

3.2. INTERPROCESS COMMUNICATION

The hearth and the most critical part of TOS is the communication mechanism between the processes. In this Section we will analyze first the model we have decided to implement and the major motivations of that choice. Then we will introduce the use of the communication primitive operations by means of few simple programming paradigms. Finally we will analize the implementation of the communication protocol using the shared memory area.

3.2.1. Model and Primitive Operations

The scheme which has been chosen is of the message passing type |CASH80|, with strict rendez-vous, explicit naming of the communicating processes and variable message length |HOAR78, CHER80|. The primitive operations take the form:

RECEIVE(Message, Sender)

and:

SEND(Message, Receiver)

This scheme implies that the two processes are brought to step to simultaneously execute the communication statements and therefore synchronization is automatically achieved as shown in Figure 3.2.

A message passing mechanism has been chosen because, although a duality was shown |LAUE79| between this scheme and communication by shared variables, message passing seems more natural for the range of applications considered and less error prone in an experimental environment in which different programming languages are used by different programmers. The syntax of the primitive operations of message passing may differ in several programming languages, but their semantics remains unchanged and this helps to reduce the error causes. Several experiences with multiprocessors have shown that communication through messages is suitable for parallel processing both at system and at user level |JONE79, OUST80|. It should be pointed out that a message passing scheme for interprocess communication does not necessarily imply any restriction in the use of global memory areas (except for those reserved for communication) by processes. It does not either mean the transformation of the multiprocessor into a network like machine.

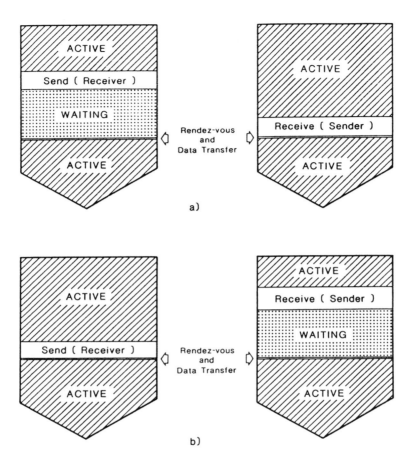

Figure 3.2 – Process synchronization by strict rendez–vous
            (a) SEND request issued before a RECEIVE
            (b) RECEIVE request issued before SEND

Moreover the implementation of a message passing scheme can be
more modular and more flexible with respect to hardware changes,
expecially in the interprocessor communication scheme.   Finally
message passing is better suited to modelling application software,
both for simulation and analytic performance evaluation methods,
because methodologies already developed for the analysis of
mainframe communication networks can be usefully exploited.
       Strict rendez–vous and therefore <u>blocking</u> primitives were
chosen because they combine synchronization and data transfer into
a single mechanism,   with no need for other functions.   Cooperating
sequential processes can be separately written as completely

sequential programs. Asynchronism is controlled by the non-deterministic communication rendez-vous and is conceptually easier to handle |MORV81|. From the implementation point of view blocking primitives allow a simpler design because no dynamic queueing of messages is required. They can be stored either at the source when the communicating processes reside in the same processor or in intermediate buffers in the common memory area when the communication takes place between processes residing on different processors. Since each process can issue only one message at a time, sizing and allocation of these areas can be done in a completely deterministic way by the off-line system generation program. This reduces the run-time software controls and overhead and avoids some deadlock problems that often arise from the bounded size of message buffers, hidden to the programmers by the semantics of a non-blocking primitive.

Unique global naming at system level and explicit addressing of processes in the communication are quite rigid constraints for application processes. On the other hand they have been found very valuable for debugging. They help to avoid side effects in communication failures, force a disciplined programming habit to the users and clearly identify interactions between processes.

Variable message length, in spite of introducing some error checking and overhead problems, was in our case an important design issue. Processes residing on different processors do not have necessarily a common address space; therefore message passing is by value and communication between processes which require a large and unpredictable amount of data to be transferred would suffer an high overhead if messages were of fixed length (packeting long messages into smaller, fixed length ones).

A third primitive operation with many-to-one semantics was introduced in order to maintain the message passing communication mechanism also for processes which provide services for several users in a non-deterministic order:

RECEIVE ANYBODY(Message)

This primitive allows a process to receive a message from any active sender in the system; the identity of the sender is made available to the receiver. In the actual implementation the selection of a sender among the waiting ones is made according to a round-robin priority scheme. Preemption by messages issued by high priority processes is foreseen as future development.

The non-determinism introduced by the fact that a process can receive a message from several different and not a-priori known sources increases the possible degree of parallelism of the system and allows the control of shared resources, like peripherals, by process structures very similar to simple Resource Managers |LAUE79| or to Proprietors |MORV81|. These processes retain the exclusive ownership of the controlled resources, at an higher level than the executive, and operate on them on behalf of the user processes which require services by means of messages. Resource handling is simplified by concentrating it into purely sequential programs and mutual exclusion among users is automatically achieved because a manager can receive one message at a time only. When an ininterruptible service lasting for more than one message is required by a user, the mutual exclusion can be extended by an explicit message protocol between the user and the Manager. In Figures 3.3 and 3.4 two examples of Manager paradigms are shown.


```
TYPE
INMessage          RECORD [Sender          WORD,
                          Information ARRAY [n] BYTE]

INTERNAL
Request            INMessage


GLOBAL

Manager            PROCESS
ENTRY

Initialize (par1, par2,...,parn)

DO
  RECEIVE ANYBODY (Request,SIZEOF Request)
  Service (Request.Information)
OD

END Manager
```


          Figure 3.3: Example of a Simple Resource Mananager

The programming language used in the examples is a pseudo-PLZ in which the notion of PROCESS has been added.

     In the first example the Manager serves the requestors on a single message basis, while in the second it defines a further set

```
TYPE
In Message RECORD    [Sender          WORD,
                      Code            WORD,
                      Information ARRAY[n] BYTE]
INTERNAL
Request        In Message
Owner          WORD


GLOBAL


Manager        PROCESS
ENTRY


Initialize (par1, par2,..., Parn)

DO

  RECEIVE ANYBODY(Request,SIZEOF Request)
  IF Request.Code = Acquire
   THEN
   Owner := Request.Sender
   DO
    RECEIVE (Owner,Request)
    IF Request.Code = Release
     THEN
     EXIT
     ELSE
     Service (Request.Information)
    FI
   OD
  FI


OD

END Manager
```

Figure 3.4: Resource Manager with Acquire/Release capabilities

of capabilities (Acquire, Release); a user must start a service with
an Acquire request and must conclude issuing a Release. During the
time between an Acquire and a Release by an user the Manager will
service request from that user only, and requests from other users
will remain pending.

3.2.2. Low Level Communication Protocol

The existence of a shared address space where data can be written
or read by any processor of the system has allowed and suggested
the use of a DFY/DAC (Data-For-You/Data-Accepted) |DOWS79|,   low
level communication protocol which does not require the use of
interprocessor   semaphores.   Semaphores   and   other   hardware
implemented communication facilities are the basis of well known
multiprocessor Operating Systems like StarOs or Medusa |JONE79,
OUST80| and   provide   very   efficient   support   of   system   and
application software,   but were not envisioned for the first TOMP
prototype.   The results of the experimentation with TOS will be the
basis for possible future extensions of the TOMP architecture with
hardware facilities for interprocess communication.
        The   protocol   is   transparent   to   processes   and   is   directly
handled on their behalf by the executive.   Communication procedures
are private ad replicated in the kernel of each processor, while the
corresponding data structures are common and allocated into the
shared memory areas.
        In this protocol two mailbox type common variables,   DFY and
DAC,   are assigned to each possible Sender/Receiver pair and the
sequence of operations is:

   - The Sender signals the availability of a message,   its address
     and length though the DFY.

   - The Receiver monitors the DFY;   when it is active,   reads the
     message,   acknowledges the receipt,   signals possible errors
     through the DAC and finally resets the DFY.

   - The Sender eventually closes the transaction resetting the DAC.

Different ways can be envisioned for the use of this protocol in a
tightly coupled multiprocessor,   according to which type of logical
unit is assumed to be Sender or a Receiver.   In fact two extreme
choices may be taken.
        According   to   the   first   one   the   primary   communication actors
are the processes and that kind of protocol provides the means for
a fully connected bidirectional communication network between them.
This   solution   allows   a   simple   and   and   regular   software
implementation,   reduces the communication overhead and is well
suited for simulation purposes.   Moreover it presents a system-wide
and instantaneous visibility of all interprocess communication for
debugging purposes.   On the other hand this solution requires a

large amount of storage area, roughly proportional to the square of the total number of processes, which is inefficiently used. In fact the maximum number of simultaneous interprocess communications is limited by our high level communication model to the total number of processes.

With the second solution the processors are the only logical communication units that are visible at protocol level and the storage requirements are reduced to the square of the number of processors. Communication between processes has to be handled locally at each processor and this requires an upper logical software structure to dispatch input messages towards internal receivers and to concentrate output messages from internal senders. It implies queue handling both at sender and receiver site. This approach was pursued in a previous multiprocessor prototype developed at the Dipartimento di Elettronica |AJMO79, CONT80|, and it showed a high communication overhead and a poor visibility for debugging purposes. Taking advantage of the blocking characteristics of the communication model with strict rendez-vous the storage requirements for a direct process-to-process protocol can be reduced.

Considering that only one message at a time can be issued by a process only one DFY and only one DAC mailboxes have to be reserved for the communication from one process towards all other processes in the system. The DFY will include an additional field for the storage of the name of the required receiver process. Therefore the two basic mailbox data types are:

DFY RECORD [Flag   BYTE,
            Receiver   WORD,
            Data Address LONG,
            Data Length   BYTE]

and:

DAC RECORD [Flag BYTE,
            Error BYTE]

and the sequence of logical protocol operations is:

  - The sender writes in its DFY the address, length of the messages and the name of the receiver. Then it monitors its DAC.

- The receiver monitors the DFY of the required sender. When it is activated and contains its own name, reads the message, resets the DFY and activates the DAC of the sender.

- The sender closes the protocol resetting its DAC.

The communication protocol is handled by the kernel of each processor on behalf of the processes residing on it. It makes an extensive use of the interprocessor interrupt mechanism supported by the global bus channel when the communication takes place between two processes which reside on different processors. In particular the processor on which the sender resides sends an interrupt to the processor on which the receiver resides, once the DFY of the sender has been written. Conversely an interrupt is sent back by the processor on which the receiver resides after the reading of the message and the activation of the sender's DAC.

The communication procedures are driven by allocation tables that contain the addresses of the mailboxes of each process and by a naming table giving the correspondence between the logical names of the processes and their location identifiers (Processor which the process is assigned to and number in the processor executive activation list). These tables are built at system generation time and allow a simple off-line reconfiguration of the system without changing either the code of the processes or the run-time kernel.

The allocation scheme that has been chosen for the communication variables has been tailored to the TOMP architecture and keeps all the data concerning the processes residing on a processor in the common memory module local to that processor. If we define as CM_i the shared memory module whose local port is directly connected to processor module PM_i and $Np[i]$ the number of processes residing on that processor, then the following data structures are assigned to CM_i :

XMIT ARRAY $N_p[i]$ DFY

REC ARRAY $N_p[i]$ DAC

corresponding to the DFY and DAC mailboxes of the processes residing on PM_i. In such a way the same protocol and the same visibility is maintained also for internal communication between processes residing on the same processor without causing overhead to processes running on other processors. In this case all the accesses to the communication mailboxes can be made though the local port of the shared memory module and do not produce traffic at global bus level.

3.3. THE EXECUTIVE

The executive is composed of a number of identical modules, the kernels, each one residing in a processor of the system and is responsible for 4 major tasks:

- System initialization

- Management and scheduling of the process execution units

- Interrupt handling

- Providing low-level monitoring functions

Except in the initialization phase, the kernels on the different processors are independent and operate basically as standard multitasking supports |KAHN78|. Therefore in the following sections we will give just an overview of their operations entering into details only for those topics which we consider typical of this implementation.

3.3.1. System Initialization

As outlined in Section 3.1.2 the system software has been developed completely on a host Z80 development system and medium speed serial connections have been used to download the software modules into the multiprocessor. Actually nearly all the TOS is resident in ROM and dowloading is used only for that part of global system data which is application dependent and for user processes. They are statically allocated to processors and reside in memory for the whole duration of the application. The dowloading procedure is executed by the user which is also responsible for placing its modules into the correct memory areas. The specifications of a system loader which would automatically execute the download procedure, have been defined but it has not yet·been implemented. By the same loading procedure each processor receives the allocation tables of the communication variables and the naming table, which have been described in Section 3.2.2 and a bootstrap table containing the number of processes allocated to the processor, and for each process:

  - The starting address,

  - The priority,

  - The USER/HANDLER specifier. This flag indicates whether the
    process is executed by the processor in SYSTEM or in
    NORMAL mode. In this last case the execution of a processor
    privileged instruction will cause a system trap. SYSTEM
    mode has to be used by processes executing I/O
    instructions.

On the basis of this information each processor initializes its
communication areas and creates the descriptors of the processes
which it will execute. Before starting the normal kernel operations
all the processors must be synchronized, that is everyone of them
must be guaranteed that all the other ones have initialized the
communication areas and are ready to handle interprocessor
interrupts. The mechanism by which this synchronization is achieved
is rather simple. Each processor sets a logical Done flag in the
common memory area after finishing internal and communication
initialization and then waits on a second Go logical flag. A master
processor waits for all Done flags to be set and then sets the Go
flags. The function of master can be assigned to any processor and
the information of which one is responsible for it at a given time
is contained in the bootstrap table of the processors. Setting and
resetting flags does not consist of simple read/write operations in
the common memory, but is based on an algorithm detecting dynamic
variable patterns in order to avoid errors or deadlocks due to
unpredictable memory patterns at startup.

3.3.2. Process Management

Process scheduling is a local activity of the kernel of each
processor and is handled with a priority scheme with up to 255
different levels. The priority of a process is set at system
generation and cannot be dynamically changed. At each priority
level scheduling follows a round-robin algorithm, that is the last
running task is given the lowest access right to the CPU. Activation
of the scheduler takes place at the occurrence of one of the
following events:

  - An Executive Call issued by a process;

- A Signal call from an interrupt procedure;

- A communication interrupt from another processor which reactivates a waiting process.

An Executive Call is either one of the previously described communication primitives, or a WAIT on an interrupt semaphore or one of the debug requests. These last are special operations, reserved to particular system processes, the Local Monitors, and will be described in detail in Section 3.3.4.

At assembly language level the Executive Calls are activated by means of the processor SYSTEM CALL instruction and the parameters are passed through processor registers. A set of procedures corresponding to these calls is available for programs written in PLZ-SYS.

3.3.3. Interrupt Handling

Interrupts are a fundamental tool in a real-time environment, where the system has to react rapidly to external stimuli. At the same time, beeing by definition asynchrounous events, they interfere in a non deterministic way with the sequential flow of execution and may cause severe side effects.

The main problem for the O.S. designer is to leave to the application programmer enough flexibility in the use of interrupts to match the real time constraints, while protecting at the same time the system from undesired interferences.

At processor level there are fundamentally 3 types of interrupt:

- System Interrupts. These are such interrupts as powerfail, memory management, interrupts, timeouts, etc.., and are directly handled by kernel routines which contain the recovery policies for these events.

  Some of these interrupts may be generated by errors in the user processes. In this case the kernel service routines automatically suspend the running process and put it into an error state, providing, at the same time, a signal to the debugging software.

- Communication Interrupts. These interprocessor interrupts, whose purposes and funtions have been described in Section 3.2.2, are directly handled by the communication procedures of the kernel.

- Peripheral Interrupts. These interrupts are the only ones which the user may require to handle directly. Complete freedom may not be allowed because it would threaten the behaviour of the whole system in case of errors and, on the other hand, enough flexibility must be granted for efficiency reasons. The compromise solution has been to keep all low level operation on peripheral devices strictly local to the processors to which the devices are physically attached and to assign all peripheral handling modules, including interrupt servicing, to one dedicated process per peripheral (Manager). Such a process is divided into two separate sections:

+ The body which handles the initialization, data formatting and communication with other processes.

+ The interrupt procedure which is responsible for the effective data transfer and synchronization with the peripheral device.

These two parts synchronize each other by the use of semaphore variables with the primitive operations of WAIT and SIGNAL. One semaphore variable is assigned to each manager at system generation and is strictly reserved to the synchronization with the corresponding interrupt routine.

The described approach, shown in Figure 3.5, is an attempt to structure the use of interrupts into three different levels with well defined interfaces and visibility among them.

At the top level there are the user processes which have no knowledge of peripherals other than the messages exchanged with the Manager. Immediately below there is the manager process level at which the visibility of the peripheral is limited to initialization operations and to to a Logical Interrupt corresponding to the semaphore operations described before. A data structure may be associated to this logical interrupt in order to exchange data with the lower level, the effective interrupt routine. This routine is executed without a complete context switching and may thus provide efficient service. Partitioning of activities between the manager and the interrupt routine is left free to the programmer; he must be aware of the fact that all operations executed at interrupt routine level are not protected by the run time virtual machine. In any case the described structure gives the possibility of shifting at Manager level most of the operations generally handled by interrupt routines, reducing the complexity of these last and hence the probability of errors.

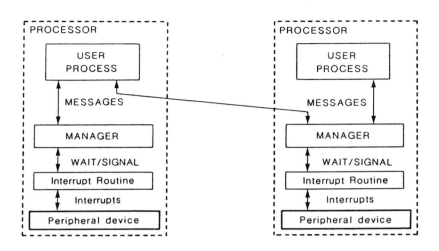

Figure 3.5: Layered peripheral operations

3.3.4. Monitoring Functions

We will describe here the functions which are provided by the
kernel for a set of system processes, the Local Monitors, in order
to allow them to interact in a controlled way with the execution of
the parallel application. These functions may be invoked explicitely
by the Local Monitor or the initiative may come from the kernel to
notify the occurrence of some event. In this last case the SIGNAL
INTERRUPT mechanism described in Section 3.3.3 is used, the kernel
playing the role of the interrupt routine.The functions provided by
the kernel are:

   - Exception Handling. When an exception either hardware
     (overflow, privileged instruction execution, etc.) or software
     (communication errors, etc.) is detected by the kernel, the
     process is suspended and the Local Monitor is notified.

   - Breakpoint handling. The kernel provides function for
     breakpointing a process either at an absolute address by
     instruction substitution or when executing a Send, Receive or
     Wait Interrupt operation. When a process encounters a
     breakpoint, it is suspended, the breakpoint disarmed and the
     Local Monitor is notified.

- Process and Communication Status. A set of functions is provided to Read/Modify the status (included the registers) of a process and to monitor and modify the status of communication variables

3.4. MONITORING AND DEBUG

3.4.1. General Architecture

For a multiprocessor programs there are two different levels of debugging. First a process may be partially be debugged as a stand alone sequential program, that is without considering the interactions with other processes. At this level well known tools like In Cicuit Emulators, Tracers, ODT's may be successfully used.

Major problems arise when debugging the cooperation of processes, in particular when executing on different processors. The lack of powerful real-time tools at this level may severely delay the acceptance and widespread use of multiprocessor systems. By powerful we mean that the visibility offered by the debugging tools to the user should not be limited only to the contents of memory, I/O and processor registers, but should be extended to process control structures and to the interprocess communication mechanism. By Real-time we mean that the debugging aids should be interactive and work concurrently with the set of application processes under exam. Their interference should be limited to the modules under test leaving the rest of the system undisturbed.

General purpose monoprocessor oriented tools give a very low visibility, in general limited to processor registers and to memory read/modify. These tools are not efficient in multiprocessor environments where a centralized control of computation (Single CPU) does not exist any more. The debugging facilities that have been developed are an experimental attempt to solve these problems and are based mainly on the following basic considerations:

- A meaningful debugging must rely on the kernel capabilities to confine, detect and categorize software errors |DENN76| and to provide a significant visibility of interprocess communication.

- In order to increase the visibility or abstraction level of a debugging tool it is necessary to tailor it to the target system to be debugged and in particular to the virtual machine supported by the executive.

- Real time and not destructive debugging is possible only if the debugger itself is not an external structure but is integrated in the run time support and runs concurrently with application programs.

- In order to debug a distributed structure the debugger itself should have distributed structure and control.

- The debugging tools of a distributed system should provide a single interaction point with the user, but with a system wide visibility.

To fulfill the above described requirements the debugger has been built as a three-layer structure which is shown in Figure 3.6.

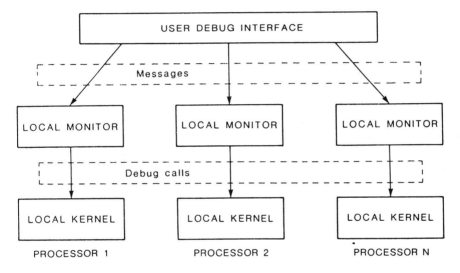

Figure 3.6 – Layered structure of the debugging environment

The first one is directly integrated into the kernel of each processor and its primary function are error detection and confinement and to to provide upper debug level with primitive operations as described in Section 3.3.4.
The second layer is composed by several identical processes (Local Monitors), one in each processor of the system, which perform in their local environment both the low level operations typical of a simple microcomputer system monitor and higher level functions, like the suspension or resumption of processes residing on the processor. Some functions are directly by the Local Monitors while the execution of others is performed by the lower level, the

kernel, on behalf of the Local Monitor. Each local Monitor sends
and receives data from/to the third layer, the Debugger Interface,
consisting of a single process running on a processor of the system.
      This process centralizes all the debugging activities of the
the system into a single operator terminal. Its functions are to edit
and parse debug commands, performing a syntactical analysis, to
format them into messages and to send them to their execution
environment (Local Monitors). It interprets the data from the Local
Monitors and displays them to the user.

3.4.2. Debugging Functions

The functions that the monitoring environment provides to the user
have been here classified according to the level of abstraction to
which they refer.

  - Register Level. This level provides the lowest level visibility
    to the user. The two available operations:

    + Memory Read/Modify

    + I/O Read/modify.

    allow him to access any memory or I/O locations. The
    operations are performed directly by the Local Monitors which
    are also responsible for protecting the system memory area
    where modifications are not allowed.

  - Process Level. By the operations provided at this level the
    user may interactively analyze the behaviour of the processes.
    The operations requested by the Debugger to the Local Monitor
    of the processor on which the process resides. The LM on its
    turn requests the execution of the operation to the processor
    kernel which is the only entity which is allowed to interact
    with the execution of a process.

    + State. This operation gives the user information about the
      state of a process and on the identity of the requested
      sender or receiver if the process is waiting for the
      execution of a communication primitive.

    + Suspend. This operation puts a process in an
      unconditionally suspended state.

+ Resume. Resumes the execution the execution of a process which has been suspended or breakpointed.

+ Breakpoint. A process can be breakpointed either at an absolute address or at a logical synchonization point, that is when executing a SEND, RECEIVE or WAIT INTERRUPT primitive operation.

+ Read/Modify Registers. This operation requires the process to be in a suspended state and allows the user to read/modify its registers.

- System Level. The operations at this level have a system wide scope and for their execution the Debugger may require the cooperation of several LMs at the same time.

+ Display of the number, the identifiers and the state of all active and suspended processes in the system.

+ Simulation of a SEND or RECEIVE operation as issued by a process towards any other in the system.

+ Simulation of a SIGNAL operation as issued by an interrupt routine.

These two last functions allow the separate debugging of different parts of the application software by the direct simulation of possible interactions between processes and peripheral devices.

3.5. UTILITIES

The utilities are system processes which run concurrently with application processes and which provide basic general purpose services and in particular an high level interface to shared system peripherals. Each utility process resides on the same processor to which the peripheral device which it serves is attached. It may nevertheless provide services for every other process in the system, independently of its location, by using the. message passing mechanism. The structure of an utility is based on that of the resource managers described in Section 3.2.1. Each utility interface with its peripherals has the two layer structure described in Section 3.3.3. In the following paragraphs we will describe briefly the operations and the services of the basic utilities which have

been installed up to now, namely the Terminal Handlers, the File
System and the Common Memory Manager.

3.5.1. Terminal Handlers

The Terminal Handlers provide the means for interfacing the user
software with the 3 video terminals of the system, each one
connected to a different processor. The basic operations of each
handler are:

- Acquire the exclusive use of the handler. This means that the
  terminal will be logically reserved for the user process which
  has issued the request. Subsequent requests from other
  processes will be queued and honored only after a Release
  from the owner process has been received.

- Release which frees the Handler from the previous Acquire.

- PutString which prints on the screen the string of ASCII
  characters which constitutes the message.

- GetString which inputs from the keyboard a terminated string
  and transfers it to the user process as a message composed by
  a sequence of ASCII characters.

A package of standard routines for simplified formatted I/O is
available for linkage with user processes. These routines handle the
task of creating, sending,receiving and decoding messages and mask
the message passing structure of the handlers.

3.5.2. File System

The File system utility provides the means for permanent data
storage. Since the the multiprocessor system does not yet have a
resident mass memory the physical storage unit is provided by the
host development system. Communication between the two systems is
provided by a medium speed serial connection. The file system is
therefore composed of two different parts, as shown in Figure 3.7.
      The first one is a process, residing on the host, which
executes operations on the mass memory on behalf of the remote file
system on the multiprocessor. The second part resides on the
multiprocessor. It has the structure of a resource manager, receives

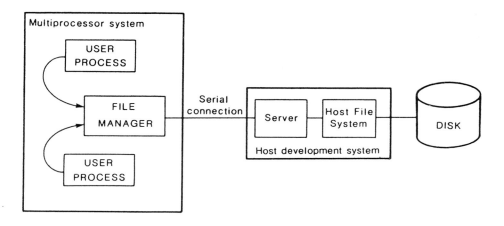

Figure 3.7: Structure of the file system

requests and data from the user processes,   formats and sends them
to the remote process for execution.   The actual implementation
allows to access a maximum of 4 files at a given time and all from
the same user process.
        The disadvantages of such a structure are clearly visible: it
is very inefficient,   slow and requires the set up of the host.
Nevertheless,   considering the constraints outlined in Section 3.1.2,
it presented also several advantages.   First it did not require the
implementation of any hardware interface.   Secondly it exploits the
capabilities of a sofware disk driver of an existing,   reliable and
debugged operating system.   Third the shared characteristics of the
storage unit allow the user to exploit all the facilities of the host
for creating data files for experimentation and for processing and
display results from the multiprocessor.

3.5.3. Common Memory Allocator

As we have already pointed out in Section 3.2.1 the message
passing mechanism does not intend to mask the shared memory
architecture of the machine,   but just to provide the means for
synchronizing the activities of the processes and for transferring a
controlled amount of information in a reliable way. When the amount

of information to be exchanged is large or when the nature of an algorithm requires a global view of a data structure, then the processes must have the possibility of a direct access to the common memory. However the use of such areas should be disciplined in order to avoid conflicts. The Common Memory Allocator is the utility which is responsible for managing the common memory area which is not reserved for system use. It provides three services to user processes:

- Allocate To request the allocation of a segment of common memory and to associate a logical identifier to it.

- Deallocate to release the memory segment corresponding to logical identifier;

- Getaddress to get the address of the memory segment associated to an identifier.

3.6. SYSTEM GENERATION

System generation is a crucial step for creating a multiprocessor application. The individual processes which comprise the parallel program are written and compiled separately and finally they must be put together into a single unit. System generation is the "glue" which links the different components and is carried on by an interactive program, called SYSGEN, running on the host development system. Its main purpose is to ease the allocation of processes to processors and of common data structures to shared memory areas, to tailor, with an automatic procedure, communication data structures to system configuration and to provide a standard documentation of the hardware and software configuration for a given experiment.

The input of the program is comprised of a number of descriptive sections relating to three different levels.

At the first one (SYSTEM) the input parameters are:

- The number of processors

- The address and length of double port memory segments assigned to processors.

At the second level (PROCESSOR) the following parameters must be specified for each processor:

- The number of processes residing on it and their identifiers (Names).

- The address, length and type (ROM/RAM) of each segment of private memory.

At the third level (PROCESS) the input parameters are:

- The entry address of the process;

- The priority;

- The address and length of any memory segment reserved for private use of the process;

- The process specifier (USER/HANDLER)

By means of these descriptions the program creates the following output files:

1. The task bootstrap table of each processor;

2. The correspondence table between the logical name of each process and the identifiers used at kernel level;

3. The allocation table of communication variables;

4. A description listing .This listing contains in a compact form all the information previously described and provides an automatic and standard documentation of the overall application system.

The SYSGEN is written in PLZ-SYS and is composed of two sequential passes. During the first it inputs data, perform controls on them, like the verification of the consistency of the allocation of processes to processors, and finally creates an intermediate file and the output listing. During the second pass it generates the object files to be loaded on the processors.

SYSGEN can be run in two different modes: Interactive and User.

The first one is used when creating a new system description and is implicit if no file name is specified when entering the program.   In this mode each parameter is explicitly requested by the program to the user.

The second mode is reserved to little and quick modifications of an already existing description and operates directly on the previously created intermediate file.   The user directly specifies parameters to be changed without reediting the whole description of the application.

3.7. A CRITICAL REVIEW

As always happens at the end of a research project when we look back at the work which has been done we realize that every problem which has been solved has also risen a set of further research issues and that for every solution which has been chosen, after its implementation we can always think of a better one.   This is even more true for this software project due to the constraints which have characterized it.

We would like here to examine in a critical way some of the design and architectural issues outlining how the constraints have limited it,   some errors which could have been avoided and future research issues which have risen from the project

Two have been the major difficulties which have been encountered in the design.   First the lack of a software development host and tools and secondly the lack of a global memory manager. The first point has made modifications and updates very expensive and has forced simplifications of the design,   expecially in the communication mechanism,   which have limited its flexibility.   We think that a reasonable software development· set for a multiprocessor should consist of at least a dedicated workstation with a large mass memory and a complete set of tools for the target machine,   comprising a compiler for an high level language, assembler,   linker and emulator.   These tools should themselves be written in a language allowing adaptation to specific requirements.

The lack of a memory manager has made very difficult to limit the propagation of errors in some code sections to other parts of the system and has consequently increased the complexity of the debug phase.   Secondly it has made difficult for application processes to take full advantage of the double port architecture of the machine due to the irregularity of the address space (processes see common memory areas at different addresses).

It is extremely important also to establish the right priorities in the development of a project if the resources are limited.   We definitely underestimated at the beginning the importance of powerful tools for automatic system generation and put the major effort of the run-time support in order to have a minimal system working in conjuction with the hardware. Then we spent a large amount of time in the tedious, repetitive and error prone of compiling and linking separately the application and system processes, loading them separately on the different processors in the correct memory areas, and until SYSGEN was available generating manually the system required tables. This time would have been shorter and better spent modifying or rewriting the primitive monoprocess linker of the host and implementing a multiprocess linker with system wide scope and a global loader utility.

In summary the implementation of the TOMP prototype software has proved to be very useful to establish the base of a better understanding of the problem related to parallel processing and has generated further research issues. In particular the actual investigation is centered on a more efficient implementation, with hardware aids, of the parallel communication and monitoring mechanisms which have been described.

3.8 REFERENCES

|AJMO79| Ajmone Marsan M., Conte G., Del Corso D. and Gregoretti F., "Architecture, Communication Procedures and Performance Evaluation of the μ^* multi-microprocessor system." In Proceedings of the 1st International Conference on Distributed Computing Systems, October, 1979.

|AJMO83| M. Ajmone Marsan, G. Balbo, G. Conte, F.Gregoretti, "Modeling Bus Contention and Memory Interference in a Multiprocessor System." IEEE Transactions on Computers, Vol.C-32(1):60-72, January, 1983.

|CASH80| Cashin P. B., "Interprocess Communication." Technical Report, Bell Northern Research, 1980.

|CHER80| Cheriton D. R., Malcom M. A., Melen L. S. and Sager, "THOTH, a portable Real Time Operating System." Communications of the ACM 22(2), February, 1980.

|JONE80| A.K.Jones,E.F.Gehringer,editors,"The Cm* Multiprocessor
              Project: A Research Review." Technical Report, Carnegie
              Mellon University, Computer Science Department, 1980.

|CONT80| Conte G.,   Gregoretti F.,   "Software development and
              debug aids for the μ^* multimicroprocessor system."
              In Proceedings of EUROMICRO Symposium. September 1980

|DENN76| Denning P. J., "Fault Tolerant Operating Systems." ACM
              Computing Surveys 8(4), December, 1976.

|DOWS79| Dowson M.,   Collins B.   and McBride B.,   "Software
              Strategy for multimicroprocessors." Microprocessors and
              Microsystems 3(8), 1979.

|MORV81| Morven   Gentleman   W.,   "Message   Passing   between
              Sequential Processes: the Reply Primitive and the
              Administrator   concept."   Software   Practice   and
              Experience 11:435-466, 1981.

|HOAR78| Hoare C. A. R., "Communicating Sequential Processes."
              Communications of the ACM 21(8), August, 1978.

|JONE80| Jones   A.K.,   Schwarz   P.,   "Experience   using
              Multiprocessor Systems - A Status Report." ACM
              Computing Surveys 12(2), June, 1980.

|JONE79| Jones A.K., Chansler R.J. Jr.,Durham I., Schwans K.,
              and Vegdahl S.R., "StarOS, a Multiprocessor Operating
              System for the support of Task Forces." In Proc. of
              the 7th Symposium of O.S. Principles. September, 1979.

|KAHN78| K.C.Kahn., "A Small Scale Operating System Foundation
              for Microprocessor Applications." Proceedings of the
              IEEE 66(2), February, 1978.

|LAUE79| Lauer H. E. and Needham R. M., "On the duality of
              Operating System structures." In D.Lanciaux (editor),
              Operating Systems: Theory and Practice, pages 371-
              384. North Holland Publishing Company, 1979.

|OUST80| Ousterhout J.K., Scelza D.A., Sindhu P.S., "Medusa: an
              experiment in distributed operating system structure."
              Communications of the ACM 23(2):92-105, February 1980.

CHAPTER 4

DESIGN OF MULTIPROCESSOR BUSES

D. Del Corso
Dipartimento di Elettronica
Politecnico di Torino
Torino – ITALY

ABSTRACT.   This Chapter describes the transfer primitives used in
parallel   bus   protocols.   The   discussion   deals   with   protocol
primitives,   signal encoding,   and synchronization techniques,   with
emphasis on the functions used in multiprocessor systems.   The
problems related with electrical interfacing are also described. A
technique to extend the basic protocol primitives is presented.

4.1. INTRODUCTION

Modular bus–based systems are composed of a set of units tied to
the same interconnection link: the bus.   All the exchanges of
information are carried out on this channel.
        All the modules are tied to the same bus,   and therefore they
communicate following a unique set of rules: the BUS PROTOCOL.
This   reference   point   forces   the   designer   towards   a   structured
approach   in   the   definition   and   development   of   the   hardware:
modules are organized in layers such as bus interface, module core,
control unit,   etc. Hardware testing becomes easier,   because the
operation of at least one side of the unit,   that is the bus
connection, can be verified using modules or tools already available
such as bus analyzers, emulators etc.
        Since all the units communicate in the same way,   the systems
are organized around a BACKPLANE BUS which is both a passive
electrical   interconnection   and   a   mechanical   support.   The
configurations of a bus–based processing system are set up by
plugging the proper physical units on the backplane. In this way a
unique library of basic modules (typically CPU, memory, and I/O of
various types) allows us to organize a system optimized for specific
purposes.

G. Conte and D. Del Corso (eds.), Multi-Microprocessor Systems for Real-Time Applications, 117–163.
© 1985 by D. Reidel Publishing Company.

Owing to the above mentioned reasons, the bus-based modularity, already widely exploited in single processor machines, is still convenient in multiprocessor systems, where the designer must face one more degree of freedom in the configuration, that is the number of processors.

Some multiprocessor machines based on single or multiple bus connection schemes have already been described in Chapter 1. From these examples we can see that in a multiprocessor structure buses are used both between a CPU and its local memory, and as global communication channels to interconnect shared resources. The requirements of these last buses are different from those found in single processor machines. Namely in the CPU-memory channel the main information flow consists of instruction fetches from the memory, while in the latter case the bus is used to exchange packets of data among the processors. Therefore any multiprocessor bus must support at least multiple mastership with an effective access mechanism.

Special protocols for processor-memory and processor-processor communications can increase the efficiency and the reliability of the information exchange. Some of these features are incorporated in recent buses, such as P896, VME, Multibus II, M3, and in the I432 structure; the detailed implementations are discussed in Chapter 5.

This Chapter analyzes the requirements and the design problems of the communication channels in a multiprocessor system. The first sections point out the logical foundations of the parallel bus protocols and define some terms used for bus specification and module design. The information exchange process involves the generation of electrical levels by means of line driving circuits, the propagation of these levels on the bus itself, and the level sensing by means of line receivers. Bus modules must be compatible also with the electrical characteristics, and the specification of a bus must describe all the electrical parameters which allow to select electrically compatible drivers, receivers, and backplanes. Finally, a technique to extend the basic protocol primitives to allow the exchange of information between more than two partners only is described.

Other general overviews on parallel buses and protocols are in |THUR72|, |THUR79|, |LEVY78|. A specific book on bus design is |DELC85|.

4.2. BASIC PROTOCOLS

4.2.1. Elementary Operations

The most simple and straightforward type of information transfer is
the POINT-TO-POINT connection.   It involves only two units,   and
moves the information in only one direction,   as shown in Figure
4.1.

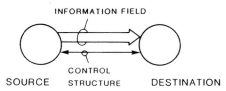

INFORMATION FIELD

CONTROL
SOURCE   STRUCTURE   DESTINATION

Fig. 4.1 - Point-to-point transfer

The modules that participate in a point-to-point transfer are:

- a SOURCE   unit   which   owns   the   information   (also   called
  TRANSMITTER);

- a DESTINATION unit,   which gets the information thanks to the
  transfer operation (also called RECEIVER).

The interconnection link is composed of two logical structures:

- an INFORMATION FIELD;

- a CONTROL STRUCTURE.

The operations on the interconnection link can be described as a
sequence of elementary ACTIONS,   such as buffer enable,   activation
of register strobe etc.   The protocol of the link specifies the
sequence of the control actions,   and the relations between them and
the information flow.
        The elementary   actions   of   the   protocol   are   encoded   into
SIGNALS which are mapped into electrical levels carried by the link
lines. The execution of an elementary operation (action) is therefore
finally expressed as a voltage level on a wire.   The words LINE or
BUS SIGNAL are used here to specify the electrical implementation of
logic variables.
        The correspondence between actions and signals is defined by
a state table.   Figure 4.2 shows two examples: an action associated

ACTION	SIGNAL	STATE
a) 1	SIG1	low-to-high or trailing edge
2	SIG1	high-to-low or leading edge

SIG 1

1   2

ACTION	SIGNAL	STATE
b) 1	SIG2	high
2	SIG3	low

SIG 2

SIG 3

1   2

Fig. 4.2 – Encoding of actions into signals:
a) Edge encoding
b) Level encoding

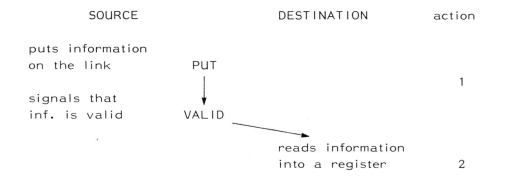

SOURCE   DESTINATION   action

puts information
on the link   PUT

                                                    1
signals that
inf. is valid   VALID

                         reads information
                         into a register   2

Fig. 4.3 – Transfer of a single item of information.

with the state of a signal ("0" or "1"),   and an action associated
with a state change ("0–1" or "1–0").

    The complete sequence of elementary actions that transfer an
item of information from the source unit to the destination is called
a CYCLE.   An example of a cycle is given in Figure 4.3.   In most
cases a link is not used for a single transfer operation,   and
therefore the sequence of operations   described above must be
followed by other similar cycles.   The subsequent cycle can start
after a time ta which includes signal settling and propagation on
the bus,   plus the time required by the destination to use (decode,
store) the information.

    The delay ta can be inserted in the sequence of protocol
actions in two ways.   A first possibility is to assume that the
information has been used within a fixed time tb   ta.   In this case
the   source   has   complete   control   of   the   timing   because,   it
synchronizes the link with the parameter tb.

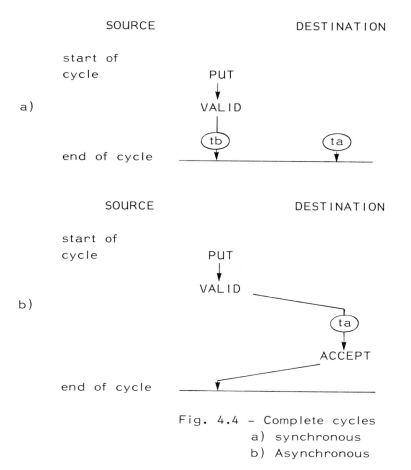

Fig. 4.4 – Complete cycles
a) synchronous
b) Asynchronous

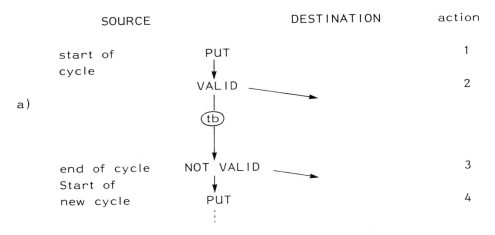

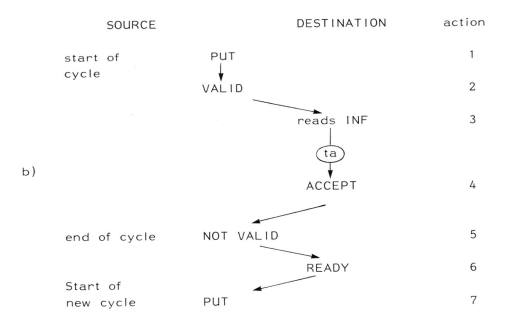

Fig. 4.5 – Closed cycles
a) Synchronous
b) Asynchronous

This transfer technique is called SYNCHRONOUS;   the corresponding action sequence is given in Figure 4.4a.

The second choice is to insert a new action,   activated by the destination,   which means that information has been used and destination is ready for a new cycle.   We shall call this action; ACCEPT this transfer technique removes fixed timing constraints from the information flow and is called ASYNCHRONOUS.   The action flow for an asynchronous cycle is given in Figure 4.4b.   This last case expresses the most elementary HANDSHAKE procedure.

To allow the chaining of identical transfer operations,   the cycle must be CLOSED,   that is the control signals must be in the same state at the beginning and at the end of each single cycle. When the actions of Figure 4.4 are encoded into electrical signals we must insert either a dummy edge or a new pair of actions to build closed cycles,   as shown in Figure 4.5.   The 4-action asynchronous cycle is called FULLY INTERLOCKED or with FULL HANDSHAKE.   Almost all the protocols described in the following use closed cycles with the 4-action handshake.   Open cycles,   that is sequences of cycles with 2-action handshake are sometimes used to get higher speed in block transfers |FAST81|, |P89683|.

4.2.2.   Types of Information Transfer Cycles

The elementary transfer cycle can also start with an information REQUEST from the destination.   The two timing choices,   synchronous or asynchronous,   are still valid,   and bring to the cycles shown respectively in Figure 4.6 and 4.7.   The source-activated and destination-activated cycles will be now analyzed focusing on the sequence of the control actions,   to merge them in a unique protocol.

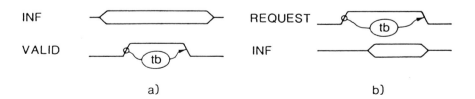

Fig. 4.6 – Synchronous cycles
a) Source-activated
b) Destination-activated

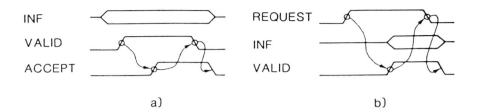

INF

VALID

ACCEPT

REQUEST

INF

VALID

a)   b)

Fig. 4.7 – Asynchronous cycles
a) Source-activated
b) Destination-activated

The module which activates the first action of the cycle will be called COMMANDER, and the other one RESPONDER. We can easily verify that the terms master/slave, already introduced in Chapter 1, have similar meaning to commander/responder. The former identify the TYPE of module:

MASTER: is able to take control of the bus and to start a transfer operation;

SLAVE: if requested by a master it can participate in a transfer operation.

The latter defines ROLES, temporarily assumed in each transaction respectively by master and slave modules:

COMMANDER: is the master which has actual control of the connection link and therefore activates the first action of the transfer operation;

RESPONDER: is the slave selected by the commander to participate in the current transfer operation.

From this definition we can see that being a master or a slave is not a variable role of bus modules, but a specific characteristic of each unit. In other words, a module is designed as a master or as a slave (or both), and can become commander or responder when it works in a system.
        Since we can define also Source/Destination roles (with respect to the direction of the information flow, as outlined in section 4.2.1), the role of a module in each cycle is completely defined by

a couple of attributes, respectively commander/ responder and source–destination. We can so define two types of cycles, as shown in Figure 4.8:

WRITE cycle: The information goes from the commander to the responder;
the COMMANDER acts as SOURCE,
the RESPONDER is the DESTINATION.

READ cycle: The information goes from the responder to the commander;
the RESPONDER acts as SOURCE,
the COMMANDER is the DESTINATION.

a)

COMMANDER   RESPONDER   COMMANDER   RESPONDER
= SOURCE   =DESTINATION   = DESTINATION   = SOURCE

b)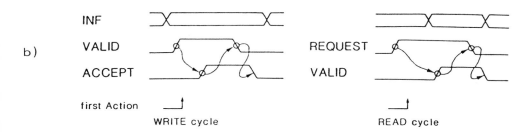

Fig. 4.8 – Read and write cycle
a) Transfer operation
b) Timing for asynchronous cycles

The protocols that allow us to exchange the source/destination roles between the commander and the responder, enabling us to perform both read and write cycles, are called READ/WRITE (RD/WR) protocols.

A description that starts from commander and responder units matches the usual architecture of processing systems better than the source/destination approach. Usually the CPU (or a DMA controller) is the commander, while memory or I/O registers are the

responders.   Data transfers are called READ or WRITE acording to
the direction of the information flow. We shall describe some choices
for the implementation of RD/WR protocols.   In these examples the
action sequence of read and write cycles is asynchronous and
follows the basic flow of Figure 4.5.

A first example of complete encoding of actions for a R/W
protocol is shown in Figure 4.9.   The commander uses the signals
WRITEPULSE (WRP) and READPULSE (RDP),   respectively in write and
read cycles,   and the responder uses an ACKNOWLEDGE (ACK) for the
handshake.   It must be pointed out that this last signal encodes
different actions in the two basic read and write cycles.   This
technique is used,   for instance,   in the P796 (Intel Multibus)
protocol |MULT79|.

	ACTION	n°	SIGNAL	STATE
WRITE cycle	VALID	1	WRP	low–to–high
	NOTVAL	3	WRP	high–to–low
	ACCEPT	2	ACK	low–to–high
	READY	4	ACK	high–to–low
READ cycle	REQUEST	5	RDP	low–to–high
	ACCEPT	7	RDP	high–to–low
	VALID	6	ACK	low–to–high
	NOTVAL	8	ACK	high–to–low

a)

b)

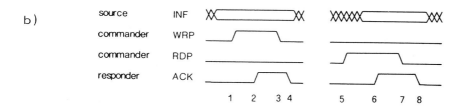

Fig. 4.9 – READUPULSE/WRITEPULSE protocol:
        a) action encoding;
        b) timing.

In order to specify the type of the cycle (Read or Write),   one can
also use a separate signal only for this purpose;   it will be

active,   for instance,   in read cycles,   and therefore will be called
READ (RD).   This technique is called ADVANCED READ or ADVANCED
WRITE,   if a write–active signal is used.   A STROBE (STB) signal
completes the encoding of the actions. RD and STB are controlled by
the commander,   and the responder uses only the ACK.   In this
protocol each action is encoded into a pair of signals: RD+STB or
RD+ACK.   The action encoding and a timing diagram of an advanced
read protocol are shown in Figure 4.10.   This technique is used,
for instance, in the VME |VMES81|.

	ACTION	n°	SIGNAL	STATE
WRITE cycle	VALID	1	STB	low–to–high
(RD low)	NOTVAL	3	STB	high–to–low
	ACCEPT	2	ACK	low–to–high
	READY	4	ACK	high–to–low
READ cycle	REQUEST	5	STB	low–to–high
(RD high)	ACCEPT	7	STB	high–to–low
	VALID	6	ACK	low–to–high
	NOTVAL	8	ACK	high–to–low

a)

b)

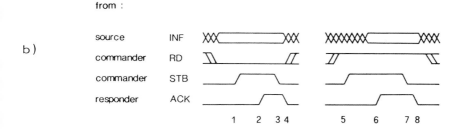

Fig. 4.10 – ADVANCED READ   protocol: The action are
        numbered as in fig. 2.7
        a) action encoding;
        b) timing.

Another example of action encoding for R/W protocol is shown in
Figure 4.11. Here the information about the direction of the transfer
is encoded in the sequencing of signal,   that is the time position of
edges.   The signals used by this protocol are VALID   (VAL) and

REQUEST (REQ), issued by the source and the destination
respectively. In this last protocol the two cycles are distinguished
only by the signal activation sequence: if VAL is active before REQ
the cycle is a write, while if REQ becomes active first the cycle is
a read. Since the read/write information is encoded into the time
domain (sequence of edges), any interface with a link that follows
this protocol must use memory elements

		ACTION	n°	SIGNAL	STATE
	WRITE cycle	VALID	1	VAL	low-to-high
		NOTVAL	3	VAL	high-to-low
a)					
		ACCEPT	2	REQ	low-to-high
		READY	4	REQ	high-to-low
	READ cycle	REQUEST	5	REQ	low-to-high
		ACCEPT	7	REQ	high-to-low
		VALID	6	VAL	low-to-high
		NOTVAL	8	VAL	high-to-low

from :

b)

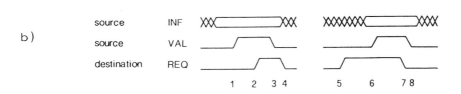

source	INF
source	VAL
destination	REQ

1 2 3 4   5   6   7 8

Fig. 4.11 – VALID/REQUEST protocol:
a) action encoding;
b) timing diagram.

(flip-flops). On the other hand, in the other cases the
read/writeinformation can be decoded from the protocol signals using
only combinatorial logic. The VAL/REQ protocol uses only two control
lines for the full handshake, at the expense of a more complex
interface.

4.2.3. Synchronization of the Action Flow

The action sequence of a protocol must run at a speed compatible with the interfaces of the modules.   In the synchronous protocols the timing is ruled only by the fixed delays built into the commander action flow.   In the asynchronous cycle the actions are interleaved in such a way that each module proceeds in the sequence only if the other partner has completed the current operation.

All the bidirectional asynchronous read/write protocols described in the previous section can be modified into the synchronous ones by removing the signals that encode the actions activated by the responder.   For the advanced read/write and the readpulse/writepulse protocol the signal ACK is deleted.   The valid/request protocol becomes synchronous by deleting REQ in write cycles, and VAL in reads. These changes are shown in Figure 4.12.

The synchronous protocol is simple but too rigid for many applications.   On the other hand the asynchronous one fits any timing requirement but uses an higher number of control lines and requires more complex interfaces on the commander and on the responder modules.   An intermediate possibility is to define a protocol that runs synchronous as default, but becomes asynchronous when there is an explicit request to change the timing.   These protocols are called SEMISYNCHRONOUS.

The action sequence of a semisynchronous protocol is derived from the synchronous one by the addition of a lock condition to the simple delay: The cycle is closed only if a blocking action from the responder is not active.   Figure 4.13 shows the action sequence and the timing of a semisynchronous write cycle;   the blocking action is encoded into the signal NOTACK.   Since NOTACK is sampled after a delay ta from VALID,   the two conditions "delay ta expired" and "NOTAC not active" are checked by a single operation.   If NOTACK is active the cycle is stopped and NOTACK continously or periodically sampled;   as soon as NOTACK is not active,   the cycle continues.

The main advantage of this technique is that in a system where all units work at the same speed the NOTACK signal is never used,   and the interface circuitry on responder modules can be very simple.   The drawback is that,   since the interlock is removed,   the source cannot be completely sure of the correctness of the operation. It must also be pointed out that the destination must activate NOTACK, when required, within a short time related to ta.

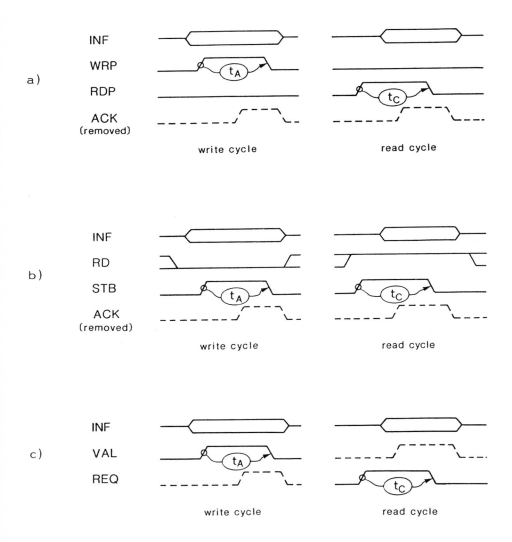

Fig. 4.12 – a) Synchronous RDP/WRP protocol.
b) Synchronous RD/STB protocol
c) Synchronous VAL/REQ protocol

This constraint creates an inconsistency in the protocol: slower
units (those that should use NOTACK),   must react within a fixed
given time: at least some part of the interface of these units must
be fast !

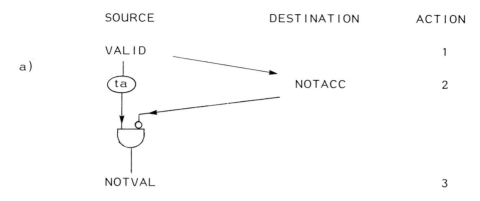

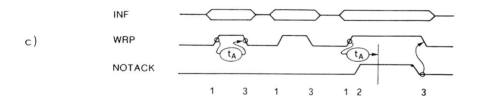

Fig. 4.13 – Semisynchronous write cycles.
a) action flow;
b) action encoding;
c) timing diagram.

The technique described above uses non-clocked (asynchronous) logic. We must however consider that in most microprocessors the link interface is integrated with the processing logic. This logic is timed by a clock, so also the sequencing of the transfer operations is related to the device clock, and therefore also the protocol actions are synchronized by the same timing signal. For this reason, when synchronous and semisynchronous protocols are

used in a clocked system, the delays are related to the clock
period rather than continously variable. In this case, inputs such
as NOTACK are periodically sampled, and the timing can be modified
only by finite steps of one clock period.

An example of complete interconnection between modules is
shown in Figure 4.14. Slave devices such as memory or I/O
interface chips have in most cases no handshake; they therefore use
synchronous protocols.

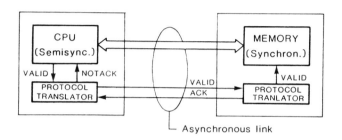

Fig. 4.14 – Example of protocol translation

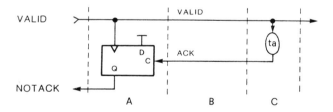

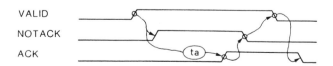

Fig. 4.15 – Protocol translation; A is semisynchronous
        B is asynchronous, C is synchronous

Master   devices,   such   as   microprocessors,   usually   exploit
semisynchronous clocked protocols to allow interfacing with slower
devices.   On a backplane a designer should preferibly use an
asynchronous protocol,   for maximum reliability and to mix boards
with different speed.   In such a system we must use PROTOCOL
TRANSLATION interfaces;   Figure 4.15 shows the translation circuits
for an advanced write protocol.

4.3. BUSED SYSTEMS

4.3.1. Channel Allocation Techniques

The bus is a single resource used by many pairs of modules to
exchange information.   As already pointed out in Chapter 1,   many
master units can be connected to the same bus; these units are able
to request the bus and to carry out independent information
transfers.   On the other hand,   the bus is a single resource which
can only carry   one information packet at a time.
        When a single resource is shared by many users,   we must
consider the possibility that many of them try to use the resource
at the same time.   The activation of many requests for a single
resource is called CONTENTION.   If many requesters,   that is many
masters,   obtain and use the bus they cause a COLLISION. In a bus
collision the protocol is corrupted,   the sequence of operationsmay
become inconsistent,   and the information transmitted is almost
always lost.
        To work properly,   only one master must have control of the
bus,   that is act as commander, at a given time. To guarantee this
condition the protocol of a multiple master bus must include the
operations to resolve contention by selecting unique commander.   The
same concept can also be applied to slave units.   Only one of them
must become responder and participate in the transfer operation.
The   process   of   selecting   one   out   of   many   slaves   is   called
ADDRESSING.
        We can see now the complete information transfer as a three-
step operation.   As shown in Figure 4.16,   the first step is the
selection of the commander among the masters which request the
bus.   As soon as the commander is identified and .has control of the
bus,   it selects one responder.   The third and last step is the data
transfer   on   the   point-to-point   commander-responder   link.   The
complete operation (arbitration + addressing + transfer) will be
called TRANSACTION.   The hierarchy of transactions,   cycles,   and
actions is show in Figure 4.17.

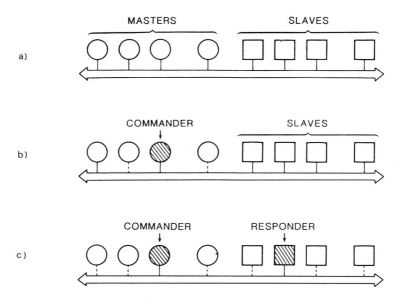

a)

b)

c)

Fig. 4.16 – Module selection process
        a) Multiple-master busse system
        b) After the arbitration cycle
        c) After the addressign cycle

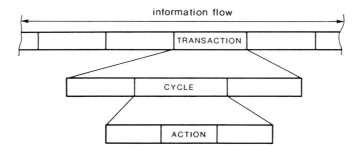

Fig. 4.17 – Hierarchy of bus operations.

The techniques for the selection of the commander in bus-based systems can be classified according to the basic mechanism used to avoid the information loss caused by collision. The choices are:

1) avoid collision by avoiding contention (token passing);

2) allow collision and avoid information loss by abort–and–retry (collision detection);

3) avoid collision by resolving the contention between requesters (arbitration).

The simplest technique to identify one out of many users is an a-priori assignement of the resource. A unique right to use is granted to each user, without considering if it actually made a resource request or not. This right can be seen as TOKEN; only the unit which owns the token can access the resource. Since each resource has its own unique token, no contention can arise and this TOKEN PASSING technique guarantees that collisions are avoided.

With the second technique each requester checks the resource status (BUSY/FREE), and starts to use it when free. Many requesters can therefore start to use the resource at the same time. The multiple use causes a collision; the collision is detected and forces all the users to release the resource and to retry later. This is called CARRIER SENSE MULTIPLE ACCESS with COLLISION DETECTION (CSMA/CD).

A third possibility to handle multiple accesses is to avoid collision by assigning the resource to only one of the requesters. All units which have a pending request for resource, participate to a contention process which selects the next user. This process is called ARBITRATION, and the subsystem which performs it is the ARBITER.

The arbitration technique, which allows contention but does not allow collision, can be seen as an intermediate possibility between the token passing (no contention, no collision), and the collision detection (both contention and collision are allowed) techniques.

4.3.2. Bus Arbitration

The logic structure of any arbitration system is shown in Figure 4.18. The arbiter translates a vector R= ri of requests into a vector G= gi of grants. To avoid collision on the resource, only one gi at a time must be active.

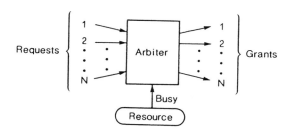

Fig. 4.18 – Logic structures of an arbiter.

When the arbiter is implemented by a single physical unit,   it is
called CENTRALIZED;   it is shown in Figure 4.19a.   The most simple
centralized arbiter is a priority encoder which activates only the
GRANT output corresponding to the highest priority REQUEST input.
An arbiter can also be made by a set of identical units,   which are
part   of   each   requester.   In   this   case   the   arbiter   has   a
DECENTRALIZED   organization   (Figure   4.19b).   If   the   complete
arbitration system is made only by   a set of identical units,   it is
also fully MODULAR.   Even with the distinction between centralized
and distributed implementations, it must be pointed out that, from a
logical point of view,   the arbiter is a single unit associated to the
shared resource.

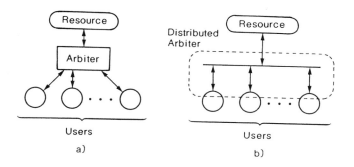

Fig. 4.19 – Arbiter organization;
a) centralized;
b) distributed.

Different techniques are used to carry the vectors R and G.   With a

centralized   arbiter,   2 N lines are required to arbitrate and select
N users.   This choice is not the most convenient for bus-based
systems,   because the request/grant lines are only point to point
connections,   organized with a star topology.   Distributed arbiters
use modular connection structures such as a bus.

A   further   degree   of   freedom,   which   identifies   different
arbitration structures,   is the PRIORITY ALGORITHM.   The priority
level assigned to any request can be fixed or variable,   according
to different rules.   When it is fixed,   the arbiter can STARVE the
users: if many high priority requests are continously coming in,
they cut off the service of low priority users. Some techniques have
been developed to avoid starvation and to make the arbiter FAIR:
they are based on dynamic change of priority levels,   or on forced
fairness.   With a fair arbiter one can define the maximum delay
which can elapse for any user from a request to the grant,   as
required by real-time applications.

A widely used technique for distributed arbitration is the
DAISY CHAIN shown in Figure 4.20.   In a daisy chain a generalized
GRANT ripples through the arbitration units of each user.   The
basic mechanism is therefore a sort of hardware token passing,
where the token is the active level of the grant line.   The daisy
chain uses only two bus lines (three or four with synchronization
signals);   the number of users can be increased at will,   the only
penalty being an increased arbitration time.   The simple chain of
Figure 4.20 can cause starvation because the priority of each user
is fixed.

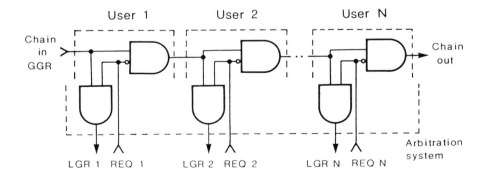

Fig. 4.20 - Distributed daisy-chain arbitration system.

With some more hardware a round robin scheduling of requests can
be implemented |ROET77|, |CIVE82|.   The benefit of round-robin

towards fixed priority is that it avoids starvation and guarantees
an upper limit to the service time.   The main weak point of the
daisy chain is that the priority level of a module depends on the
physical position in the backplane.   Modules must have fixed slots,
and there is no possibility of changing the priority levels at run
time.   Moreover,   a single fault or a missing module can lock the
chain or cause the activation of multiple grants.
        A solution well suited for modular bus-based systems is the
distributed arbiter with self-selection priority networks |TAUB76|,
|TAUB84|. With this technique each unit participates in a contention
process which is based on a single-bit collision   detection on the
bus lines.   The unit with highest priority gets the bus for the next
transfer operation.   N priority lines allows one to arbitrate 2^N
requesters,   with a delay roughly proportional to N |TAUB82|,
|DELC84|.   The priority can be dynamically changed,   and the
identity of the current master is visible from the bus.   By means of
suitable fairness algorithms,   it is also possible to guarantee that
each requesting module will get the bus within a known time for
any combination of requests from other modules |CIVE83|.

4.3.3. The Distributed Self-selection Arbiter

This technique is described in detail because it is used with slight
changes in some recent buses |S10079|, |FAST81|, |M3BU81|,
|P89683|. A self-selection distributed arbiter consists of one unit on
each master,   connected by a set of priority lines and some control
lines, as shown in Figure 4.21.

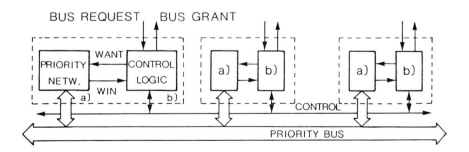

Fig. 4.21 - Distributed self-selection arbitration system.
              a) priority network;
              b) control logic

The elementary arbitration units are composed of:

- a self-selection priority network;

- a control block.

Each control block receives a BUS REQUEST from the bus access
circuitry of the master module.   Then it handles the arbitration
procedure by sending appropriate commands to the priority network
and to the arbitration control lines.   The self-selection network
carries out the process that identifies the highest priority module.
This last unit wins the arbitration and issues the BUS GRANT to its
master, which becomes the commander of the bus for the current
transaction.
        The block diagram of a self selection network is shown in
Figure 4.22. The network is divided into three blocks:

- Vector Enabling (VE) gates: they gate the local priority vector
  IP to the bus drivers:

- Open Collector buffers (OC):   they put the gated priority code GP
  on the priority bus lines BP*;

- Bit Comparators (BC):   they check if each bit of the module code
  IP matches the logic level on the corresponding BP* line.

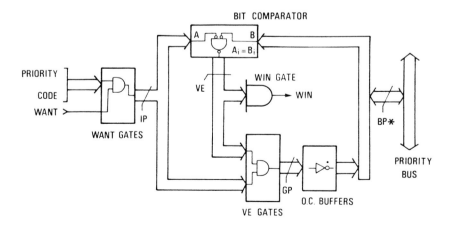

Fig. 4.22 - Block diagram of a self-selection network

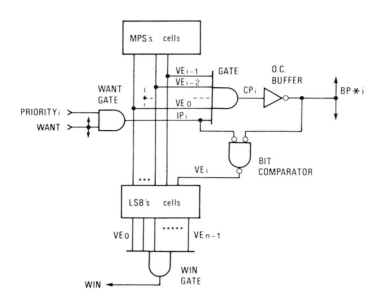

Fig. 4.23 – Self–selection priority network;
a) block diagram of a network;
b) priority circuits

The arrangement of bit comparators, BP* drivers, and enabling gates for each line of the priority bus is shown in Figure 4.23. If a bit comparator detects a mismatch on bit i, it disables all the GPh of weight h < i. When two or more networks are connected to the same priority bus through inverting open collector drivers, each line carries the logic AND of GP*. The truth table of the priority logic is shown in Figure 4.24.

The logic equations of the network are:

$$GPi = IPi \qquad \text{if all } VEh = 1 \qquad \text{for } h > i$$

$$GPi = 0 \qquad \text{if } IPi = 0 \quad \text{or if any } VEh = 0 \text{ for } h > i$$

When the vector on the bus matches the priority code of the module, all VEs are true and the output WIN is activated.

We shall now describe, as an example, the behaviour of two 4–bit networks A and B with codes respectively 0101 and 0011, connected to the same priority bus. The complete diagram of a 4–bit selection network is shown in Figure 4.25.

	IPi	CPi	BPi*	VEi
a)	0	0	0	0
b)	0	0	1	1
c)	1	1	1	1
d)	1	1	1	–

Fig. 4.24 – Truth table of self-selection logic.
    a) BPi* is forced to 0 by another priority
       network with IPi=1;
    b) IPi=0 in all networks;
    c) BPi* goes to 0 because IPi=1 in this network;
       in other networks it can be either 0 or 1;
    d) This case is not allowed, because BPi* is
       forced to 0 when at least one Ipi=1.

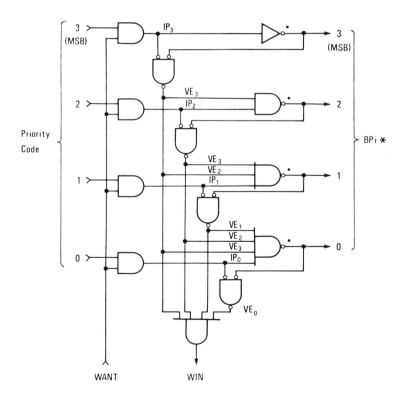

Fig. 4.25 – Complete diagram of a 1-bit self selection network.

For the codes specified above,   there is agreement on MSB because
$IP3a=IP3b=0$; $BP3^*=1$, therefore both VE3a and VE3b are high. On bit
2,   the comparator of network b detects a disagreement ($IP2a=1$,
therefore $BP2^*=0$,   bit $IP2b=0$),   and disables the drivers of lower
priority lines ($CP2b=CP1b=CP0b=0$).   At this point the lines $BP1^*$ and
$BP0^*$ are controlled by network a only,   and the final result of the
process   is   that   no   bit   comparator   of   this   network   senses   a
disagreement (all $VEa=1$).   This condition is decoded and activates
the WIN output.   The state of the networks after the contention is
shown in Figure 4.26.

bit	Network A			Network B			bus
	IP	CP	VE	IP	CP	VE	BP*
3(MSB)	0	0	1	0	0	1	1
2	1	1	1	0	0	0	0
1	0	0	1	1	0	1	1
0	1	1	1	1	0	1	0
	WIN = 1			WIN = 0			

Fig. 4.26 – States of the two networks after the contention

It must be pointed out that the self selection process involves the
propagation of logic changes through many gates.   The WIN output
must be sampled after a   WANT request with a delay to allow for
gate propagation and   set–up on bus lines.   This delay is called
CONTENTION   time.   The   control   logic   must   take   care   of   this
constraint   to   generate   a   correct   GRANT   towards   the   master.
Techniques to evaluate the contention time are given in |TAUB82|
and |DELC84|.
     The signals   exchanged   with   the   priority   network   are   only
WANT   and   WIN.   The   sequence   of   actions   shown   in   Figure   4.27
defines the operation of the control logic as follows:

1) The master activates BUS REQUEST;

2) As soon as a new arbitration cycle can start,   the control logic
   activates WANT.   This happens for all modules with pending bus
   requests;

3) The   priority   networks   with   WANT   =1   start   the   self   selection
   process.   After this step no other module can activate WANT or

interfere in the self-selection process;

4) After a suitable delay Ta (greater than the contention time),
   WIN is tested.   If true,   BUS GRANT goes high and the master
   becomes the commander for the current bus transaction.   If WIN
   is false,   GRANT is kept inactive and the master waits.   BUS
   REQUEST stays active and the master will participate in the next
   arbitration cycle.

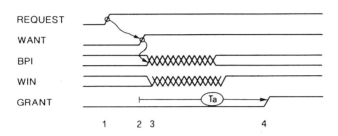

Fig. 4.27 - Arbitration cycle

This sequence must be synchronized with other bus operations. While
the priority network has a "universal" structure, the implementation
of the control block is related to the protocol of the specific bus.
        If the sequence of the arbitration action is timed by a system
clock signal,   the arbitration procedure is called SYNCHRONOUS.
When the actions are sequenced by a handshake chain, and there is
no unique time reference,   the arbitration is ASYNCHRONOUS.   The
first technique is used in M3BUS |M3BU81|,   the second in P896
|P89683|.

4.4. ELECTRICAL BEHAVIOUR OF BACKPLANE LINES

4.4.1. Definition of Signal Levels

To guarantee electrical compatibility of logic modules one must first
check how logic states are mapped into voltage levels. Owing to the
spread of device characteristcs, the two states of a binary variable
correspond to two ranges of voltage levels.   Logic devices exhibit
compatible output and input levels if "0" and "1" output levels are
also always seen respectively as "0" and "1" input levels.   This
compatibility must include some margin for noise,   voltage drops on

signal paths, etc., as shown in Figure 4.28.

Any device tied to a line draws current from it, and the line driver must supply enough current for at least all the receivers connected to the line, plus the leakage current of other disabled drivers, and for the termination networks. From the output and input Figures we can derive the maximum number of receivers that can be driven by a single transmitter, as shown in Figure 4.29.

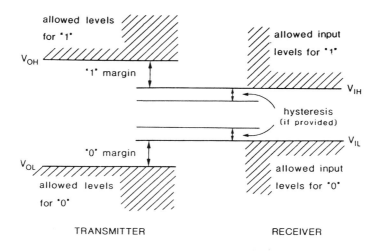

Fig. 4.28 – Definition of logic levels for a bus line.

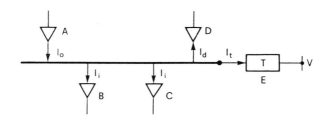

Fig. 4.29 – Loads on a bus line:
              A) active driver;
              B,C) receivers;
              D) disabled driver;
              E) termination network.

More details and the actual values of voltage and currents levels can be found for any family of driver/receiver in the documentation of manufacturers.   We shall now look   at the behaviour of logic circuits connected to a bus,   to point out the specific issues which must be considered in this situation.

4.4.2. Transmission Line Effects

A   bus   is   usually   implemented   on   a   printed-circuit   board (backplane) as a set of parallel tracks.   For the usual dielectric constant of printed circuit substrates ($\varepsilon r=5$) the propagation delay on a track is about 6 ns/m. This gives, in the full-rack length (50 cm) a travelling time comparable with the transition time of TTL circuits,   therefore the system cannot be considered as being made up of lumped elements only,   and the tracks must be treated as trasmission lines |DEFA70|, |BALA84|.
     A   trasmission   line   is   described   by   the   characteristic impedance Zo which depends on the physical structure,   as shown in Figure 4.30 |KAUP67|.

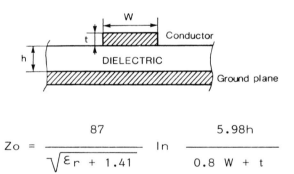

$$Zo = \frac{87}{\sqrt{\varepsilon r + 1.41}} \; \ln \; \frac{5.98h}{0.8\ W + t}$$

Fig. 4.30 - Characteristic impedance of a backplane line

In a backplane with line width and spacing suitable for DIN 41612 connectors, the typical values of Zo lies   in the range 50-130 ohms. When the output of the line driver changes,   the actual voltage step at the output can be computed from the equivalent circuit of Figure 4.31a.   Since the driver is a non-linear device, its quiescent output resistance   Rd   must   be   graphically   derived   from   the   output characteristic, as shown in Figure 4.31b.
     The voltage step propagates on the line,   and is partially reflected by every impedance mismatch.   A simple graphic technique to draw the waveform on the line is Bergeron's method |FAIR78|, |DELC85|.   The track is terminated at the ends in some way (open-

circuit,   pull-up   resistors,   etc.);   drivers   and   receivers   are
connected almost randomly along the line itself,   and   this also
changes the characteristic impedance and causes mismatches.

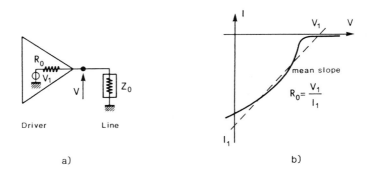

a)   b)

Fig. 4.31 - a) Simplified equivalent circuit of the
                      driver-line-termination system.
                   b) Output characteristic of TTL drivers.

If the input impedance of line receivers is much higher than the
line impedance Zo,   the most significant discontinuities occur at the
ends of the line,   and depend on the value Rt of the termination
networks. If Rt is greater than Zo, the reflected wave increases the
voltage on the line, and the successive reflections add each other.
        For these reasons the voltage on the line reaches a value
which is interpreted as a "1" by receivers only after a settling
time ts which depends on Zo,   Rd, and Rt. Rd is a characteristic of
the driver and varies between different TTL families.   The effects
mentioned above limit the data rate in a backplane. A level change
is sensed by a receiver only when the input voltage crosses the
threshold.   As shown in Figure 4.32,   this may happen at the first
step or later, depending on the Rd/Zo ratio |BALA84|.
        Two usual termination circuits are given in Figure 4.33. A
correct termination of each line can eliminate most of the reflections
but requires drivers capable of supplying the currents required by
the terminations.   Let us consider for instance a line with Zo = 100
ohm,   terminated at both ends with 100 ohm resistors, tied to a + 3
V supply.   To drive the line under a 0.5 V logic "0" threshold, the
transmitter must sink 25 mA from each termination (50 mA total),
plus Iol of the receivers and leakage of other devices. The designer
must also consider that,   even when the termination is correct,   the
backplane track is far from being a pure stripline because the

Connectors introduce mismatches and drivers and receivers are non-
linear devices.

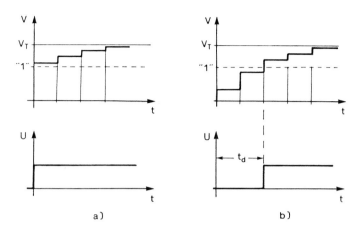

Fig. 4.32 – Delay caused by insufficient driver current
a) Good driving; threshold is crossed
without reflection
b) The threshold is crossed at second reflection.

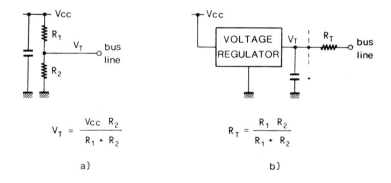

Fig. 4.33 – Terminator of bus lines
a) passive termination;
b) active termination.

The overall effect is that the line impedance cannot be exactly
specified in the actual operating environment, and it is almost
impossible to define fixed rules for the design of matched interface
circuitry in these conditions. If the worst-case reflections are

impossible to define fixed rules for the design of matched interface
circuitry in these conditions. If the worst-case reflections are
known by means of analytical or experimental techniques, the
settling time can be derived from the voltage waveforms and thein
put threshold of the receivers.

4.4.3 Crosstalk

For the whole lenght of the backplane the bus lines run spaced
closely together; the capacitive and inductive coupling can modify
the voltage levels on the lines, as shown in Figure 4.34. This
effect is called CROSSTALK, and may become an important cause of
errors in the information transfer.

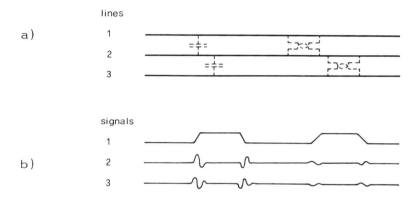

Fig. 4.34 - Crosstalk between backplane lines.
                a) equivalent circuit of parasitics.
                b) effects on bus signals: the induced
                   noise is roughly proportional to
                   edge slope ($\Delta V/\Delta t$)

The crosstalk on a backplane depends on line spacing, routing,
ground location, termination networks, and driver output impedance.
Crosstalk becomes a problem when the voltage induced on a quiet
line by adjacent lines is high enough to produce a false logic
reading at the receiver. The worst case occurs when all lines but
one switch together in such a way to add the induced noise on the
steady line. This type of error is extremely difficult to fix in a
running system, therefore one must use proper techniques in the
electrical design and in the layout in order to limit the crosstalk

level within the threshold margins of the driver/receiver pair.

The two causes of crosstalk are electrical and magnetic coupling. The noise induced by capacitive coupling is proportional to dV/dt of the transmitter and to the input resistance of the receiver. In our case the receiver is a bus line, defined by Rd, Zo, Rt. To reduce this crosstalk one should limit the slope of signal edges, and lower the impedance of line drivers and terminators. Capacitive interferences can also be reduced by means of interleaved ground tracks and ground planes which act as electrostatic shields.

The inductively coupled noise is proportional to dI/dt at the trasmitter and to the input admittance of the receiver. In a bus it can be reduced by raising the line impedance and by limiting the current of drivers, that is by raising the value of termination resistors. Inductive coupling is not affected by ground shields.

There is no unique design choice which minimizes the overall crosstalk. A technique to control the crosstalk is the limitation of the slope of waveforms edges; it becomes more effective if used together with input integration. As shown in Figure 4.35 this last technique allows one to eliminate short spikes caused by reflection, crosstalk, or by the wired–OR glitches |GUST83|.

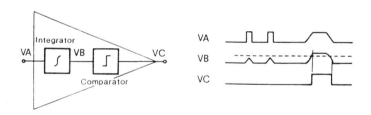

Fig. 4.35 – Controlled–slope transmitter and integrating receiver to avoid crosstalk noise

Controlled output slope and input integration are combined in the devices called TRAPEZOIDAL TRANSCEIVERS |BALA81|, |BALA84|. The high noise immunity obtained with this kind of active protection is achieved at the expense of the time loss during slow edges and during input integration, which limit the rate of change of bus electrical signals.

In a backplane it is common practice to use passive protection techniques, such as capacitive shielding of tracks. An example of shielded layout is shown in Figure 4.36.

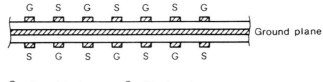

G = Ground track   S = Signal track

Fig. 4.36 – Backplane layout to reduce crosstalk. Signal
tracks (S) are alternated with ground tracks (G);
a ground plane further reduces cross–coupling
between adjacent lines.

Long or high–speed backplanes use twisted signal–ground pairs or
differential drivers to minimize the noise radiated and picked up.
Other techniques rely on the fact that crosstalk is synchronous with
signal switching.   Since the interferences are localized in time,
they can be avoided if signal lines are sampled when they are
settled,   as shown in Figure 4.37.   This technique again introduces
delays,   and can be used only for lines which carry static
information,   such as DATA or ADDRESS signals.   For this reason in
some cases the electrical specifications of edge–active strobe lines
are different from those of static information lines |VMES81|,
|P89683|.

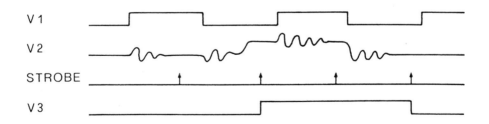

Fig. 3.37 – Delayed sampling to avoid crosstalk.
V1: Disturbing waveform
V2: Signal with crosstalk
V3: Resampled signal

4.4.4 Protocol Speed

We can now evaluate the speed of a protocol, being aware of the
main factors which limit the throughput of a bus defined in the
previous section. The technique presented here allows us to find the
maximum cycle rate which can be afforded by a protocol, while
keeping the correct sequence of elementary actions.

Let us consider, for instance, the transfer of a single data
item with full handshake shown in Figure 4.38. At the first step
the transmitter puts data on the bus and signals that the data is
valid. As soon as VALID is sensed, the receiver initiates a process,
such as generating the strobe pulse for a register or for a memory,
in order to store or to use in other ways the incoming data. After
this, or concurrently if the store process is fast enough, the
receiver activates ACKNOWLEDGE and continues the sequence of
protocol actions.

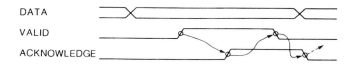

DATA

VALID

ACKNOWLEDGE

Fig. 4.38 – Basic 4-edge handshake.

In order to describe the actual behaviour of electrical signals
in a backplane we must consider that the transmitter and the
receiver are placed at different locations. Electrical levels must
propagate and settle, and for this reason a signal is sensed by a
receiver only after a time tp, from the activation by the trasmitter.
This delay includes edge slope, signal propagation and settling,
and input filtering (if used). The final effect is that the timing at
the receiver is different from the timing at the transmitter, because
the actual value of the delays depends on the relative position of
the two units and on the parameters Rd, Zo, Rt. Even when all
the above parameters are specified, the delay tp is affected by an
uncertainty due to the spread of device characteristics, and to the
various possible positions of the boards in the backplane. The total
delay is:

$$tp = tp' + Dtp$$

where Dtp is the uncertainty in the propagation and settling time.

The designer knows only the nominal delay tp' and the upper limit
of Dtp.   A properly designed interface must work in the full range
of tp.   An example of the effects of Dtp on the time relations of two
signals is shown in Figure 4.39.   Dtp causes a time skew,   and the
transmitter data set-up time tsus is reduced at the receiver to

$$tsud = tsus - Dtp$$

For this reason,   to guarantee a set-up time greater than zero at
the receiver,   VALID must be delayed by at least a Dtp. This delay
is called (transmitter) DESKEW TIME.

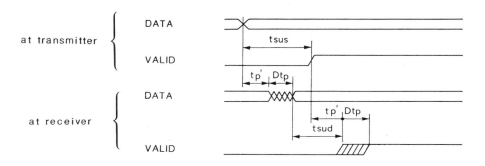

Fig. 4.39 - Skew   and   propagation.   tb'   and   tb are the
              propagation and   the   skew   time   of data.   tp is
              the propagation time of VALID.   tsus and tsud are
              the   set-up   time   respectively   at   the transmitter
              and at the receiver.

We now have all the elements to start an analysis of the protocol,
considering the effects of signal propagation to find the actual cyle
speed.   The complete sequence of the asynchronous handshake shown
in Figure 4.40 includes a detailed view of all delays and will be
used as one reference example.   The time starts when the trasmitter
outputs the data,   but only when VALID is actually active can the
receiver process the data .   The delay from VALID (at the receiver)
to "information used" depends on the internal structure of the
destination and is not considered in this analysis.   The time tc
required to complete the transfer cycle,   provided that (t7 + t9)>(t8 +
t10), is:

$$tc = t1 + t3 + t4 + t6 + t7 + t9 + t11 + t12$$

The inverse of tc is the speed S of the protocol in cycles per second. If P is the parallelism of the data path, the throughput B of a bus is:

B = S * P

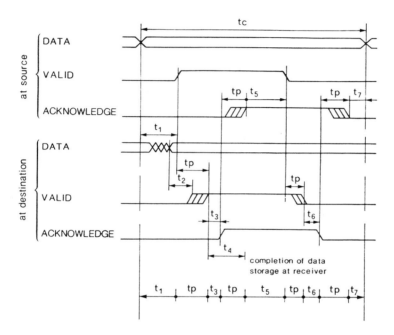

Fig. 4.40 – Combined effects of propagation and skew on bus signals timing.

4.5. PROTOCOL EXTENSION

4.5.1. The Enable/Disable Technique

We shall now analyze how the transfer procedures can be extended to allow the exchange of information between three or more units. The modules involved in the information transfer may have different speed requirements, but the protocol must always guarantee the asynchronous handshake. This condition is fulfilled if the following two prerequisites are met:

- any module can detect if any other module has activated an action on the bus (does at least one partner respond to the operation ?);
- any module, looking at the bus, can detect if all partners agree on activating an action (do all partners agree on performing the next operation ?).

These two conditions are satisfied if one can obtain both the OR and the AND of all the activation signals, as shown in Figure 4.41a. Using the wired-logic techniques, these logic operations are performed on the bus itself, as shown in Figure 4.41b. This approach is well suited to bussed systems, because it is modular with respect to the number of input variables: to add one input to this AND/OR logic function, one must only tie a new driver to the bus line.

a)

b)

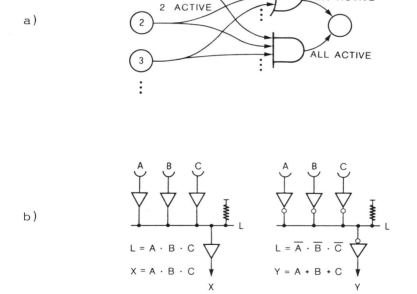

Fig. 4.41 - N-partner protocol.
        a) Control signals
        b) Wired logic with TTL open collector gates.

The first test (does any partner respond to the operation request?) requires ORing of signals. With standard TTL open collector drivers these signals must be encoded into active-low bus lines. The second test (do all partners agree ?) requires ANDing, that is active-high signals on the bus lines. Therefore a pair of bus signals is needed; it will be called here the ENABLE/DISABLE (E/D) pair. This E/D concept has been already used in a debug environment |DELC79|, and has been further exploited to extend the primitives of an information exchange protocol |DELC82|.

In a protocol using the E/D technique, the activation of an action X is triggered by a double condition:

-    an X Enable (X-EN) signal is active;

-    an X Disable (X-DIS) signal is not active.

For an action X controlled by an E/D pair, with the wired logic shown in Figure 4.41b, every unit can read from the bus:

- the first activation of the action, looking at the active low X-EN
   line (at least one module has activated the action, and so X-EN
   is low);

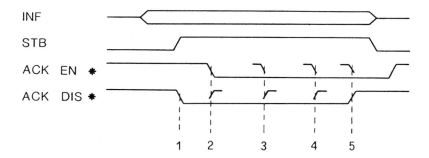

Fig. 4.42 – The E/D pair technique applied to the ACCEPT action.
             The sequence for acknowledge from N partners is:
     1)   ACKDIS* must go low before ACKEN* becomes active;
     2)   The fastest destination accepts INF; ACKEN* goes low
           (active), and ACKDIS* stays low;
  3,4)   Intermediate speed destinations accept INF; both
           ACKEN* and ACKDIS are not affected;
     5)   The slowest destination accepts INF and deactivates
           ACKDIS*. This line is no longer active and the source
           considers INF accepted.

– the last activation of the same action, looking at the active low
X–DIS (all disable signals are not active and so X–DIS is high).

This method can be applied for instance to the ACCEPT action, that
is to the acknowledge signal.   It allows fully handshaked
information transfer towards multiple destinations (broadcast),   as
shown in Figure 4.42.
        The same technique can be used also to control other actions
of the protocol.   For instance,   Figure 4.43 shows how to delay the
VALID action.

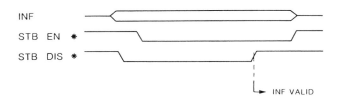

Fig. 4.43 – The E/D pair applied to the VALID action.

The combination of the delayed strobe (E/D pair for VALID action)
with an E/D pair for the INF buffer enable shown in Figure 4.44
allows one to replace the information put on the bus by the source
with the information coming from another module.   In this case the
buffer enable action is controlled by the two signals:

BUF–EN   : internal to the module, and

BUF–DIS : external;   this bus signal is controlled by the unit that
                 performs the substitution.

It must be pointed out that if X–DIS is never activated,   the action
X is ruled by X–EN only;   so X–DIS is actually an optional signal
and also any module that handles X–DIS is optional. The units that
control the optional protocol extension signal of an E/D pair are
here called SUPERVISORs.   A system which supports N–partner
information exchanges,   works correctly even when no supervisor is
connected to the bus and all transactions involve only two modules.
        Up to this point,   the E/D pair technique has been used to
define the logical structure of the basic mechanism that allows the
interaction with an information transfer.   This method makes it
possible to implement new protocol primitives such as:

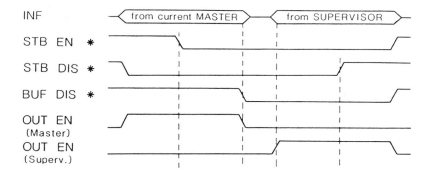

Fig. 4.44 – Replacement of the information on the bus.
1)   INF from the source is valid but not accepted by the destination because STBDIS* is active.
2)   INF buffer of the master is disabled (BUFDIS* active)
3)   INF buffer of the supervisor is enabled.
4)   INF from the supervisor is valid and can be accepted by the destination (STBDIS* not active).

- asynchronous handshake between N partners: it allows one to change the timing of a bus operation;

- source buffer enable controlled from the bus: it allows the replacement of information;

- change of operation type: it allows one for instance to change a write cycle into a dummy operation, for memory write protect.

The next section will show how these primitives can be used for modular upgrade of processing systems by means of supervisory units.

4.5.2.   Bus Supervisors

As shown in Figure 4.46, we can distinguish two types of interaction in an information exchange.

MONITORING: a third unit simply looks at the transaction, without
                   modifying it in any way;   this operation can be
                   considered a 1-Source to 2-Destination transfer.

INTERVENTION: the   third   unit   takes   an   active   part   in   the
                   transaction and can modify some parameters of the
                   control   structure   or   of   the   information   being
                   transferred.

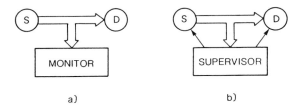

a)   b)

Fig. 4.45 - Observation (a) and intervention
(b) in an information transfer.

In the first case the monitoring unit must capture on the fly the
desired   information   from   the   current   transaction.   A   general
discussion on hardware and software monitoring techniques is in
|PLAT80|.   Oscilloscopes,   logic analyzers and more sophisticated
execution monitors,   such   as   the   Snoop   developed   for   Fastbus
|WALZ80|, and the Fast-Bus-Inspector developed for M3BUS |DOTT81|,
belong to this category.
        We   shall   here   concentrate   mainly   on   the   second   type   of
interaction.   The units capable of intervention must support an N-
partner protocol,   therefore they are supervisors.   It is now clear
that this term does not define a single module,   but a family of
units with different functions. Various levels of intervention can be
defined, depending on the actual interactions of the supervisor with
the transaction.
        The   lowest   level   is   the   simple   observation,   with   the
possibility of modifying the speed of the information transfer to fit
the timing requirements of the monitor. A stronger interaction occurs
when   the   information   being   transferred   is   changed   for   the
replacement of the information.   An intermediate intervention level
is,   for instance,   the inhibition of an erroneous update of the
destination register (write protect).   Supervisors with the above
defined capabilities can be used for many purposes   both in the
debug environment and at run time.
        The monitoring of information exchange allows one tracing and

breakpointing (via the interrupt structure of the processor). These operations can be performed without affecting the real time behaviour of the observed system if the supervisor is fast enough.

The intervention capability of the supervisor allows one to delay and to synchronize the observed process to perform more complex and time-consuming operations, such as multiple table search, symbolic tracing, etc. The replacement of information can be used in the debug phase to switch memory banks of a development system into the memory of the target machine or to reallocate memory accesses.

Some of these features are also supported by In Circuit Emulators (ICE). The supervisor approach removes some problems that arise from the use of an ICE, such as those due to differences in timing and electrical characteristics between the emulator probe and the actual device. The supervisor is connected to the bus; since bus lines are buffered, terminated, and have well defined electrical characteristics a connection with them should not influence the behavoiur of the system. Moreover, the supervisor hardware is defined acording to the system bus rather than from the CPU specifications. The same hardware unit can therefore support monitoring and intervention on many interface sections, provided that they use the same protocol.

This feature is extremely usefull in multiprocessor systems where the information visible on a CPU interface does not represent the system state |DEMA85|.

To give an example of complex supervisor interaction, we shall now show how the information replacement technique shown in Figure 4.44 can be used to implement memory management functions. To fully understand the difference from more usual structures, we shall divide the memory management process, into the following steps:

- a consistency check on the information (address range, operation type, privilege level etc). If an error is found, the processor is notified via the interrupt or trap mechanism, and the operation is aborted.

- a replacement of the logical address issued by the processor with a physical address for the memory.

The former operation belongs to the monitoring category; the latter to the replacement one; both can be performed by the proper supervisor. Compared with the other techniques shown in Figure 4.46, a memory management supervisor is fully modular and, owing to the greater design freedom, can be more powerful and versatile.

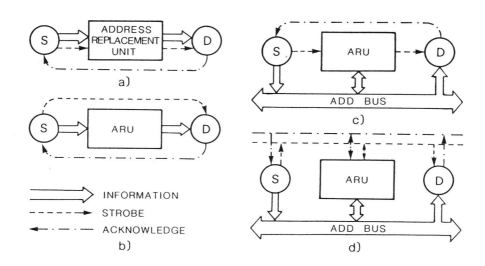

Fig. 4.46 – Techniques for memory management.
> a) The Address Replacement Unit (AWRU) is
>   connected between the source and the
>   destination. Both INF and control signals are
>   modified by the ARU.
> b) The ARU affects INF only; controls must take
>   care off additional delays (Z8001 approach).
> c) The ARU replaces INF and delays STROBE (used
>   in N16032).
> d) Fully modular ARU (memory management by
>   means of supervisors.

In the example of this section we used separate E/D pairs for each
operation, but it must be pointed out that the E/D technique does
not require a complete duplication of control lines. Since the
operation type is usually already specified by one of the E/D
signals, the other one can be multiplexed. An example of complete
extended protocol with only two additional lines is given in
|DELC82|.

4.6. REFERENCES

|BALA81| Balakrishnan, R.V., "DS 3662 – the bus optimizer", National Semiconductor Application Note 259, April 1981.

|BALA84| Balakrishnan, R.V., "The Proposed IEEE 896 Futurebus: A Solution to the Bus Driving Problem", IEEE Micro, vol. 4, No 4, August 1984

|CIVE82| Civera, P. et al, "The u* Project: an experience with a multimicroprocessor system", IEEE Micro, May 1982.

|CIVE83| Civera, P. et al, "An integrated self-selection arbiter", EUROMICRO 83 Proc., Madrid, October 1983.

|DEFA70| De Falco, J.A., "Reflection and crosstalk in logic circuit interconnections", IEEE Spectrum, July 1970.

|DELC79| Del Corso, D., "A test technique for microprocessor-based machines", Alta Frequenza, February 1979.

|DELC82| Del Corso, D., Maddaleno, F., "Extension of bus protocols: a technique for modular upgrade of processing systems", EUROMICRO 82 Proceedings, Haifa, September 1982.

|DELC84| Del Corso, D., Verrua, L., "Contention delay in distributed priority networks", Microprocessing and Microprogramming, n. 1, January 1984.

|DELC85| Del Corso, D., et al, "Microcomputer Buses and Links" Academic Press, 1985.

|DEMA85| Del Corso, D., et al., "Distribution of monitoring and control functions in multiprocessor systems", internal report 1984, to be published.

|DOTT81| Dotti, D., et al., "Caratteristiche funzionali e strutturali dell'FBI (Fast Bus Inspector)", AICA-81, Pavia, September 1981.

|FAIR78| "Fairchild TTL data book", 1978.

|FAST81| U.S. NIM Committe,   "FASTBUS tentative specification",
              August 1981.

|GUST83| Gustavson,   D.,   Theus,   J.,   "Wire-OR   logic   on
              transmission lines", IEEE Micro, june 1983.

|KAUP67| Kaupp,   H.R.,   "Characteristics of microstrip lines",
              IEEE Trans. on Computers, April 1967.

|LEVY78| Levy,   J.V.,   "Buses,   the   skeleton   of   computer
              structures",   Computer   Enginnering: a   DEC   view   of
              Hardware System Design, Digital Press, 1978.

|MULT79| "Intel MULTIBUS specification", 1979.
              - a summary of MULTIBUS specifications is in Chapter
              5.2 of this book.

|M3BU81| Del Corso,   D., Duchi, G., "M3BUS: System specification
              for   high   performance   multimicroprocessor   machines",
              BIAS 1981 Proceedings, Milano, October 1981.
              - a summary of M3 specifications is in Chapter 5.5 of
              this book.

|PLAT81| Plattner,   B.,   Nievergelt,   J.,   "Monitoring Program
              Execution: a survey", IEEE Computer, November 1981.

|P89684| IEEE   P896   Committee   "P896   Specification",   November
              1983.
              - a summary of P896 specifications is in Chapter 5.4
              of this book.

|ROEN77| Roethlisberger,   H.,   "A standard bus for multiprocessor
              architecture",   EUROMICRO 77 Proceedings,   Amsterdam,
              October 1977.

|S10079| Elmquist,   K.A. et al, "Standard Specification for S-100
              Bus Interface Devices", IEEE Computer, July 1979.

|TAUB76| Taub,   D.M., "Contention resolving circuits for computer
              interrupt systems",   Proc.   IEE,   num.   9,   September
              1976.

|TAUB82| Taub, D.M., "Worst–case arbitration time in S100–type
              computer bus systems", Electronics Letters, n. 18,
              September 1982.

|TAUB84| Taub, M., "Arbitration and Control Acquisition in the
              Proposed IEEE 896 Futurebus", IEEE Micro, vol. 4, No
              4, August 1984.

|THUR72| Thurber, K.J., et al, "A systematic approach to the
              design   of   digital   bussing   structures",   AFIPS
              Conference Proc., Vol. 41, 1972.

|THUR79| Thurber, K.J., Masson, G.M., "Distributed Processor
              Communication Architectures", Lexington Books, 1979.

|VMES81| "VMEBUS Specification Manual", 1981.
              – a summary of VME specifications is in Chapter 5.3
              of this book.

|WALZ80| Walz, H.V., Downing, R., "Fastbus Snoop Diagnostic
              Module",   1980 Nuclear Science and Nuclear power
              System Symp., Orlando, Florida, November 1980.

CHAPTER 5

SOME EXAMPLES OF MULTIPROCESSOR BUSES

P. Civera, D. Del Corso, F. Maddaleno
Dipartimento di Elettronica
Politecnico di Torino
Torino, ITALY

ABSTRACT   This chapter describes four buses suitable to be used in multiprocessor systems: MULTIBUS,   VME,   P896,   and M3.   The electrical specifications,   the protocol,   and the key features are discussed for each bus.   The functions   related to multiprocessor operations are emphasized.   An analysis of timing requirements and of transfer speed is also carried out.   The M3BUS standard is described more in detail.

5.1. INTRODUCTION

The purpose of this chapter is to make a comparison between four backplane buses able to support at least 16-bit microprocessor systems. They have been selected to illustrate examples of different design philosophies, such as multiplexed/non-multiplexed structure, centralized/decentralized arbitration, etc.. All these buses are "standard", that is they are defined in a specification document which is accepted by many manufacturers and, in some cases (Multibus, P896), also by organizations of international standards.

Multibus is one of the first buses designed from the beginning as a standard; it is supported by a wide number of manufacturers and represents a classic reference point for designers. VME has been introduced to support more powerful processors and quickly gained wide market approval P896 is a "future" project not currently available on the market, which pushes advanced solutions and is optimized for 32-bit processors. M3BUS is trademark of the Italian National Research Council (CNR), and is used in Italy by large and small companies in process automation and control. The authors have direct design experience on this standard, and for this reason it is described in more detail.

165

G. Conte and D. Del Corso (eds.), Multi-Microprocessor Systems for Real-Time Applications, 165–224.
© 1985 by D. Reidel Publishing Company.

To allow easier comparison, all buses are here described following the same bottom-up structure: mechanical and electrical characteristics first, protocol and special functions later. The names of the signals are those used in the original specification documents, but the general definitions, schemes, timing diagrams, follow the conventions already introduced in Chapter 4.

5.2. THE MULTIBUS BACKPLANE

5.2.1. History and Main Features

The "Multibus" backplane |MULT79| was originally developed by Intel to support the family of 8080/86 products, which includes 8-bit and 16-bit microcomputer boards, memories and I/O (both digital and analog). An early version of Multibus was submitted for standardization to the IEEE MSC (Microcomputer Standard Commettee) with the name of P796.

This bus is not multiplexed, has a data path width of 16 bits, but can also support 8-bit microprocessors and memories using the byte swap technique. The memory address space is of 16 Mbytes and the I/O space is of 64 kports.

In order to set up multiprocessor systems, more than one master is allowed on the same bus segment. To handle bus control in these cases, two different arbitration mechanisms are provided. Indivisible operations (even on dual port memories) are also supported.

5.2.2. Physical and Electrical Specifications

Multibus uses two direct connectors, named P1 and P2, for each board. P1 is an 86-pin (2x43) board edge direct connector. It carries the most important signals and lines (power supply, address, data, control).

The auxiliary connector P2 is a 60-pin (2x30) board edge connector. It carries auxiliary signals, such as address extension, auxiliary power supply, power fail, and has some reserved pins, both bused or not. The user can use non-bused pins, providing a suitable motherboard wiring: in case of non-bused wiring, the different positions on the backplane are no longer equivalent and the boards using these signals are position dependent.

The Multibus card dimensions and the numbering of connector pins are shown in Figure 5.1.

The backplane distributes power supplies to the logic, digital and analog interfaces, and battery powered backup, as specified in Table 5.1.

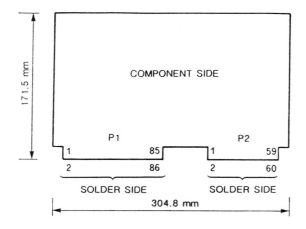

Fig. 5.1 - Multibus board and pin numbering

Table 5.1
Bus Power Specifications

Voltage	Use	Connect.	Tolerance	Ripple
5 V	Logic	P1	5%	50 mV
12 V	Logic	P1	5%	50 mV
-12 V	Logic	P1	5%	50 mV
15 V	Analog	P2	3%	10 mV
-15 V	Analog	P2	3%	10 mV
5 V	Back up	P2	5%	50 mV
12 V	Back up	P2	5%	50 mV
-12 V	Back up	P2	5%	50 mV
-5 V	BacK up	P2	5%	50 mV

The voltage levels specified for transmitters and receivers are given in Figure 5.2. Each line must be terminated on the backplane with

CHAPTER 5

a pull up resistor or a resistive voltage divider between +5V and ground, in order to reduce the reflections and the settling time on the bus. The driver and receiver currents are different for each class of signals. Some examples are given in Table 5.11.

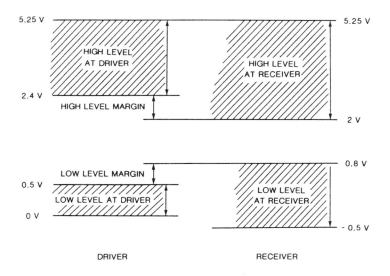

DRIVER   RECEIVER

Fig. 5.2 – Multibus electrical levels

Table 5.11
Examples of specifications for bus drivers, receivers and terminations.

| Signal | Driver Currents | | Receiver Currents | | Terminations |
	I_{OH}	I_{OL}	I_{IH}	I_{IL}	P.up P.down (kohms)
DAT0/–DATF/	–2 mA	16 mA	125 μA	–.8 mA	2.2
ADR0/–ADR17/	–2 mA	16 mA	125 μA	–.8 mA	2.2
MRDC/ MWTC/					
IORC/ IOWC/	–2 mA	32 mA	125 μA	–2 mA	1
XACK/	–2 mA	32 mA	125 μA	–2 mA	.510
BCLK/	–3 mA	48 mA	125 μA	–2 mA	.220 .330
BREQ/	–.2 mA	10 mA	50 μA	–2 mA	Not Required

The rise and fall time for the signals   driven by totem pole or three-state drivers should be 10 ns; for the open collector lines this condition applies only to the fall time.   The bus propagation delay must be less than 3 ns, from any edge connector   to any other one, when the line is driven by a 74S20.   The specifications for set-up and hold time, ringing and crosstalk are given in Figure 5.3.   In any case, at receiver, the ringing on command lines and the crosstalk must not exceed the noise immunity levels.

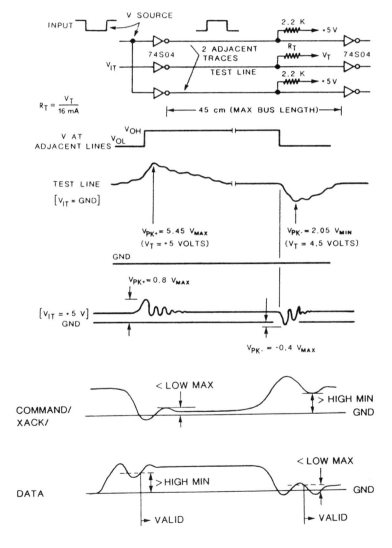

Fig. 5.3 - Ringing and crosstalk specification

5.2.3. The Information Transfer Protocol

Each information transfer on Multibus can be decomposed in two sequential operations:

- Multibus exchange
- Data transfer

The first operation permits the selection of the master which has control of the bus; in the second part this master (now the commander) performs the addressing of the slave and the information transfer. Due to the fact that Multibus has independent arbitration and transfer structures, these two operations, for two different masters, can occur in the same time slot.

Multibus provides two different mechanisms for the ￜbus exchange: a daisy-chain or a centralized arbiter. The following bus signals are used for the two arbitration techniques:

Bus CLocK (BCLK/): synchronizes all the bus exchange logic on its fall edge.

Bus BUSY (BUSY/): indicates that the bus is currently in use; it is controlled by the commander.

Bus PRiority iN (BPRN/): indicates that no master of higher priority is requesting the Multibus control. In the daisy-chain scheme, this signal comes from the preceding master in the chain. In the parallel arbitration scheme, this signal is the grant coming from the central arbiter.

Common Bus ReQuest (CBRQ/): indicates that a master is waiting for the bus, which is currently controlled by another master.

The trailing slash (/) following a signal name means that it is active low.

Common Bus Request is used to save time in bus exchange: if the commander at the end of its data transfer finds that no other master is requesting the bus, it retains the bus for a possible next information transfer. This saves the bus exchange overhead when the same commander keeps the bus for many consecutive transfers. By this means, in a single master system, the bus exchange operation does not take place at all.

In a system using a daisy-chain scheme for the priority resolution, each master has an output, BPRO/ (Bus PRiority Output),

which indicates to the following master in the chain that a higher
priority master is requesting the bus. The daisy-chain connection
is shown in Figure 5.4a.

With the parallel priority resolution schemes, each master signals
its bus request to the arbiter by activating its Bus REQuest line
(BREQ/). The arbiter grants the winner using the Bus PRiority iN
(BPRN/) signal. This priority resolution scheme is shown in Figure
5.4b.

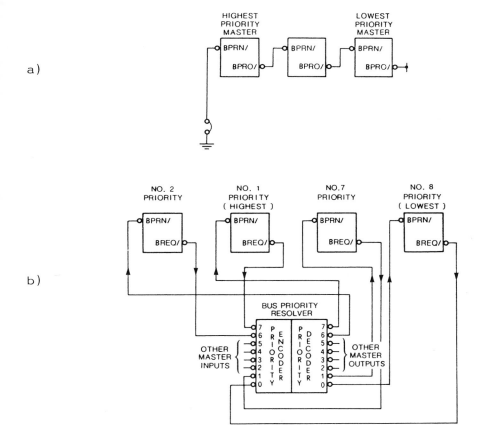

Fig. 5.4 - Arbitration schemes
    a) daisy-chain connection;
    b) parallel connection.

When a master has control of the bus, it can begin the data
transfer operation. In Multibus four command lines are provided,
thus allowing four different kinds of data transfer:

- MRDC/ :   Memory ReaD Command;
- IORC/ :   I/O Read Command;
- MWTC/ :   Memory WriTe Command;
- IOWC/ :   I/O Write Command.

In a read operation, the commander puts the memory or I/O address (using 24 or 16 bits respectively) on the address lines. After the set-up time, the read command MRDC/ or IORC/ is issued by the commander. When the responder has prepared the requested information, it puts it on the bus data lines, and activates the acknowledge signal XACK/ at the same time. When the commander receives the acknowledge signal, it accepts the information, removes the read command and, after the hold time, it also removes the address. Finally, the responder removes the acknowledge signal and the data. The sequence of signals in a read operation is shown in Figure 5.5a. The difference between a memory and I/O read operation consists in the number of address bits used and in the different command line activated.

        In a write operation, the master puts the address (24 or 16 bits, depending if it is a memory or I/O operation) and the data on the bus at the same time. After the set-up time, the master issues the write command (MWTC/ or IOWC/). The selected slave accepts the data and activates the acknowledge line. The master then removes the write command, and after the hold time, it also removes the address and data. Finally, the responder removes the acknowledge signal, thus closing the data transfer. The signal evolution of a write operation is shown in Figure 5.5b.

        Multibus can perform data transfers of 8 or 16 bits at a time: the two signals, Bus High ENable (BHEN/) and the lowest significant address bit A0/, control the type of transfer. Multibus defines the even byte staying on the 8 lowest (less significant) data bits of the bus, and the odd byte on the highest (most significant) 8 data bits. Four different formats of data transfer are possible:

        - 16-bit word transfer;
        - even byte transfer (always on the low byte);
        - odd byte transfer on the high data byte;
        - odd byte transfer on the low data byte.

This last kind of transfer has been defined for 8-bit masters (always connected to the low data byte) working with 16-bit memories. It requires a byte swap buffer in order to send the high byte of the memory on the low byte data path on the bus.

a)

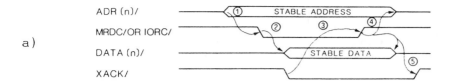

1) Address set-up time;
2) Responder access time;
3) Commander accept time;
4) Address hold time;
5) Responder release time.

b)

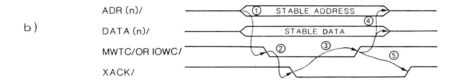

1) Address and data set-up time;
2) Responder accept time;
3) Commander release time;
4) Address and data hold time;
5) Responder release time.

Fig. 5.5 – Data transfer operations
a) Read operation timing
b) Write operation timing

The Bus High Enable signal indicates if the upper byte of the data bus is in use, and the less significant address bit A0/ specifies if the transfer involves an odd or even byte. A word transfer is coded as an even address with high byte enabled. The different formats and data paths of data transfer are shown in Figure 5.6.

Indivisible operations (typically read-modify-write) are possible on Multibus. To perform such an operation, the system must guarantee to a commander that no other master has access to the memory between the indivisible read and write operations.

The mutual exclusion among masters, connected on the same bus segment, is obtained using the Bus Busy signal. In fact, the commander which begins an indivisible operation does not release the bus control until the end of the operation. On the contrary, a processor physically residing on the same card of amemory, as

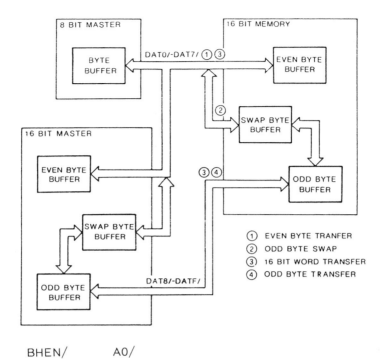

	BHEN/	A0/	
1)	1	1	Even byte transfer
2)	1	0	Odd byte transfer on low bus byte
3)	0	1	Word transfer
4)	0	0	Odd byte transfer on high bus byte

Fig. 5.6 – Data tansfer formats

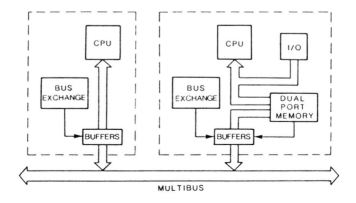

Fig. 5.7 – System with a dual port memory

shown   in Figure 5.7,   does not use the bus to access to its
memory.   In this case,   another signal is necessary to prevent the
local processor from accessing the dual port memory during the
indivisible operations performed by a master connected to the bus.
This bus signal is named LOCK/;   it   is activated by the
commander performing an indivisible operation and is sensed by the
dual port memory interface.

5.2.4. Special Features

Two   special   features   of   Multibus   are   discussed: inhibit   and
interrupt.

    Multibus provides a mechanism for inserting, on the same bus,
a ROM memory at the same address of a RAM,   without causing
conflicts.   Moreover,   a ROM can also override another ROM memory.
These capabilities are used,   for instance,   for a ROM containing a
bootstrap   routine   which   overrides   the   normal   system   RAM   at   the
start up.   As another example,   a ROM memory containing diagnostic
programs,   when activated,   can inhibit the system ROM at the same
address.

    These   features   are   obtained   using   2   bus   lines: INH1/
(INHibit 1),   which prevents the RAM memories from responding, and
INH2/ (INHibit 2) which also inhibits the ROM memories.   The effect
of Inhibit lines is not defined during write operations.   An example
of connection for these lines is given in Figure 5.8.

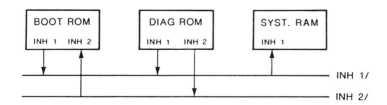

Fig. 5.8 – Use of inhibit lines

Multibus supports two classes of interrupts,   called respectively Non
Bus Vectored and Bus Vectored.   The Non Bus Vectored interrupts are
handled by special circuitry on the master card and do not require
vector   transfers   on   the   bus.   The   slave   which   generates   the
interrupt   request   activates   one   of   the   8   interrupt   request   lines
provided on the   bus.   The interrupt vector is generated on the
master board by the interrupt controller,   and the interrupt request
is cleared by the master with an explicit operation on the slave

(e.g. write or read operation): no interrupt acknowledge signal is
used in this case. Figure 5.9 shows an example of Non Bus Vectored
structure and the related timing.

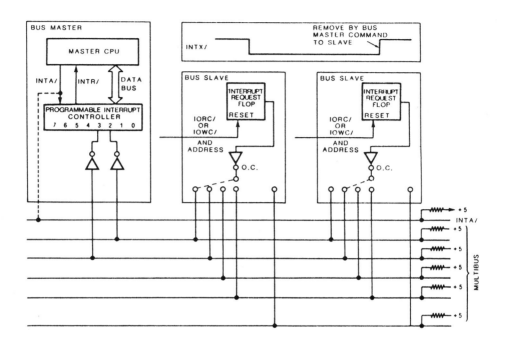

Fig. 5.9 - Non Bus Vectored interrupt structure

The Bus Vectored interrupts require the transfer of an interrupt
vector from the slave to the master on the bus. An INTerrupt
Acknowledge signal (INTA/) is used to synchronize this transfer. A
slave module requests an interrupt by activating one of the 8
interrupt request lines. The master sends a first pulse on the
interrupt acknowledge line in order to inhibit other requests. Then
the master sends the code of the interrupting device with highest
priority on the address lines, and activates the interrupt
acknowledge line a second time to request the vector. This way, it
is possible to sequentially handle many requests sent at the same
time. The interrupting device which recognizes its priority code on
the address lines sends its interrupt vector on the data lines and
validates it with the acknowledge signal. The master accepts the
vector and deactivates the INTA/ line. Finally, the slave

deactivates the acknowledge signal and the interrupt request. An implementation example of bus vectored interrupt and the related timing is shown in Figure 5.10.

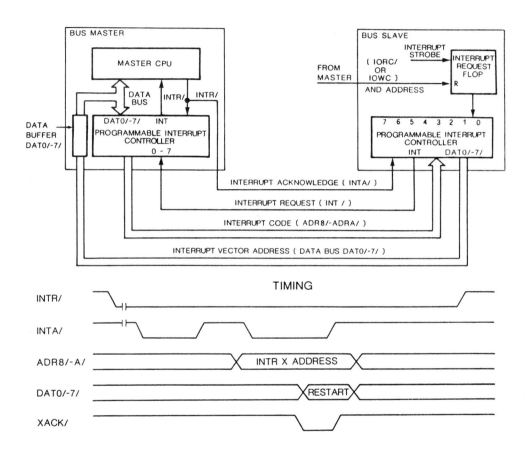

Fig. 5.10 – Bus Vectored interrupt structure and timing

If the vector to be passed from the slave to the master is larger than 16 bits, two information transfers are necessary. In this case the master activates the INTA/ signal a third time and the slave sends the second part of the vector. Both vectored and non-vectored interrupts can be used in a system at the same time, but the Bus Vectored interrupt can be only of a single type (with one or two data transfers to move the vector).

5.2.5. Timing and Pinout

The pinout of Multibus is shown in Figure 5.11.

For simplicity, only the timing of information transfer is discussed in this section.

In addressing and write operations, the information must be stable at destination with a set-up time of 50 ns whit respect to the valid command. This means that the source must also consider the skew time introduced by propagation. The information hold time at destination must be of 50 ns minimum.

During the read operations, the responder can activate the acknowledge signal as soon as the data are in the bus. The deskew must be done by the commander.

In any case, the responder must release the bus lines which it controls (acknowledge and data during the read operations) within 65 ns from the end of the command.

COMPONENT SIDE				CIRCUIT SIDE	
1	GND		2		GND
3	+5 Batt.		4		+5 Batt.
5	Res. not bused		6		EEVPP
7	-5 Batt.		8		-5 Batt.
9	Res. not bused		10		Res. not bused
11	+12 Batt.		12		+12 Batt.
13	PFSR/		14		Res. not bused
15	-12 Batt.		16		-12 Batt.
17	PFSN/		18		ACLO/
19	PFIN/		20		MPRO/
21	GND		22		GND
23	+15		24		+15
25	-15		26		-15
27	PAR1/		28		HALT/
29	PAR2/		30		WAIT/
31	PLC		32		ALE
33	Res. not bused		34		Res. not bused
35	"		36		BD RESET/
37	"		38		AUX RESET/
39	"		40		Res. not bused
41	Res. bused		42		Res. bused
43	"		44		"
45	"		46		"
47	"		48		"
49	"		50		"
51	"		52		"
53	"		54		"
55	ADR16/		56		ADR17/
57	ADR14/		58		ADR15/
59	Res. bused		60		Res. bused

Fig. 5.11a – Pinout of MULTIBUS connector P2

COMPONENT SIDE				CIRCUIT SIDE	
1	GND		2	GND	
3	+5		4	+5	
5	+5		6	+5	
7	+12		8	+12	
9	Res. bused		10	Res. bused	
11	GND		12	GND	
13	BCLK/		14	INIT/	
15	BPRN/		16	BPRO/	
17	BUSY/		18	BREQ/	
19	MRDC/		20	MWTC/	
21	IORC/		22	IOWC/	
23	XACK/		24	INH1/	
25	LOCK/		26	INH2/	
27	BHEN/		28	AD10/	
29	CBRQ/		30	AD11/	
31	CCLK/		32	AD12/	
33	INTA/		34	AD13/	
35	INT6/		36	INT7/	
37	INT4/		38	INT5/	
39	INT2/		40	INT3/	
41	INT0/		42	INT1/	
43	ADRE/		44	ADRF/	
45	ADRC/		46	ADRD/	
47	ADRA/		48	ADRB/	
49	ADR8/		50	ADR9/	
51	ADR6/		52	ADR7/	
53	ADR4/		54	ADR5/	
55	ADR2/		56	ADR3	
57	ADR0/		58	ADR1	
59	DATE/		60	DATF/	
61	DATC/		62	DATD/	
63	DATA/		64	DATB/	
65	DAT8/		66	DAT9/	
67	DAT6/		68	DAT7/	
69	DAT4/		70	DAT5/	
71	DAT2/		72	DAT3/	
73	DAT0/		74	DAT1/	
75	GND		76	GND	
77	Res. bused		78	Res. bused	
79	−12		80	−12	
81	+5		82	+5	
83	+5		84	+5	
85	GND		86	GND	

Fig. 5.11b − Pinout of MULTIBUS connector P1

5.3. THE VME BACKPLANE BUS

5.3.1. History and Main Features

The VME bus standard |VMEB81| was developed in 1981 by Mostek, Motorola and Signetics/Philips to support the new generation of 16 and 32-bit microcomputers such as the 68000 Motorola microprocessor family.   The design goal was a high speed and high performance bus with powerful interrupt management and multiprocessor capability.

The bus is fully parallel with asynchronous handshaken protocol.   Three different levels of bus complexity are described by the specification: standard,   reduced,   and extended, with different width for the address and the data paths.   VME compatible boards must define the compliance level.   The standard level has 16 Mbytes of address space and 16-bit data path,   the extended type has 4 Gbytes of address space and 32-bit data path,   while 64 kbytes of address space and 16-bit data path are defined for the reduced configuration.

More than one master can be supported by the same bus in order to obtain multiprocessor systems. The arbitration mechanism is a multi-level daisy-chain.

The bus signals are mapped on a 96-pins indirect connector. A second auxiliary connector for user definable I/O signals is also provided.   The extended configuration uses part of the auxiliary connector to carry the exceeding address and data signals.

5.3.2. Physical and Electrical Specifications

VME uses standard double Euro-size boards and two 96-pin DIN 41612 connectors with females mounted on the backplane.   The board size is derived from the IEC standards for 19" rack |IEC249, IEC297,   IEC603| as in P896 and M3 bus standards;   Figure 5.12 shows the mechanical outlines.   Connector P1 carries all the bus signals.   In the standard and reduced configuration also single size boards with P1 only are allowed. Connector P2 carries extended bus signals, if any, and user definable I/O signals.

In this way,   different positions on the backplane are no longer equivalent and boards using connector P2 for I/O must go into a fixed position.

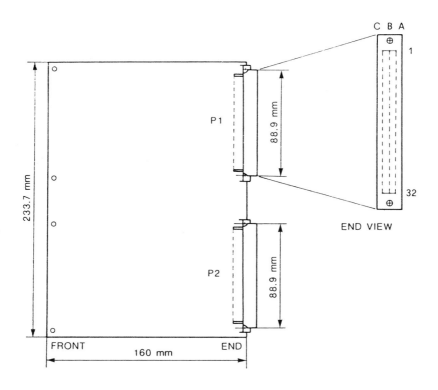

Fig. 5.12 - VME eurocard designation and pin numbering

The backplane distributes the main power supply and battery backup supply digital and analog circuits.

Up to 22 boards can be connected to the same VME segment; the first (A0) and the last (A21) connectors are reserved for the termination resistors, and the first available connector (A1) accomodates the bus arbiter and the reset/power-on module.

The signal levels are TTL standard, as specified in Figure 5.13. Depending on the function required three-state, open collector, and totem pole drivers are used as transmitters. For bused signals a 48 mA current sink capability is required, therefore standard TTL-LS devices can be used only for daisy-chain lines. The bus lines are terminated with a 330/470 ohms resistor network or equivalent circuit.

The effective unloaded characteristic impedance of each line should be lower than 100 ohms. The total capacitance of a board for each signal line must not exceed 25 pF.

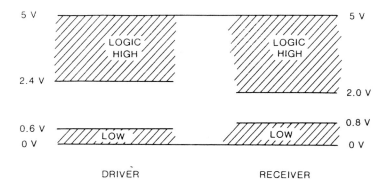

Fig. 5.13 – VME electrical levels

5.3.3. The Information Transfer Protocol

Each information transfer on VME consists of two sequential cycles:

    – Arbitration
    – Addressing and data transfer

Arbitration allows a master to gain control of the bus, the second
cycle selects the slave and transfers the information.
        The first cycle is used only on multimaster systems when a
new master requests the bus to become commander. The arbitration
cycle is always performed before the addressing and the data
transfer cycle using a reserved set of lines.
        Addressing and data transfer are carried out concurrently by
the commander because of the non-multiplexed structure.
        The bus arbitration is based on the daisy-chain scheme. As
shown in Figure 5.14, four daisy-chains for four different levels
of priority ripple through the masters.
        With the daisy-chain lines and other bused lines (BBSY*,
BCLR*) three different arbitration policies are offered: single
daisy-chain, fixed priority and rotating priority. They are
selectable on the arbiter and master boards. In the single daisy-
chain the master's fixed priority is defined by its geographic
position with respect to the other masters and to the arbiter. The
timing relationship of the arbitration signals is represented in
Figure 5.15.

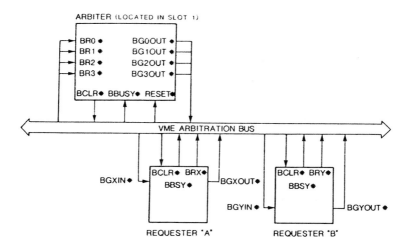

Fig. 5.14 – Block diagram of the arbitration structure, 2 requesters.

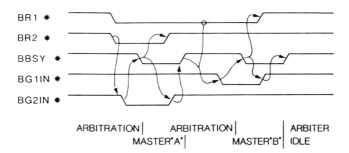

Fig. 5.15 – Arbitration cycles with 2 requesters on 2 different levels

The bus slot A1 is reserved for the arbiter circuitry and consequently the bus slot A2 is the highest priority slot in the bus. The signals dedicated to the arbitration operation are:

BR* <0..3> Bus Request, daisy-chain signals (output from master);

BG* <0..3> Bus Grant, daisy-chain signals (input to master);

BBSY* Bus BuSY, active when the commander is using the bus;

BCLR* Bus CLeaR, used in fixed priority only, informs the current master that a higher priority request is pending.

The address/data cycle uses independent sets of lines for address, data, and control. In the VME this cycle is called Data Transfer Bus (DTB). The transfer protocol is asynchronous and it handles a single responder. The control signals used by the commander are:

AS*   Address Strobe;
DS0*   Data Strobe 0: when active enables the odd byte data transfer operation;
DS1*   Data Strobe 1: when active enables the even byte data transfer operation;
LWORD*   Long WORD: when active specifies a 32-bit data transfer (extended mode only);
WRITE*   specifies the direction of the data transfer.

The information transmitted by the commander consists of the address and 6 address modifier bits, which specify the addressing type, the addressing range (standard or extended) and the privilege level.

The handshake signal DTACK* is activated by the addressed slave to acknowledge the transfer. BERR* is activated instead of DTACK* by the time-out circuitry when the responder does not exist or after unallowed or erroneus data transfer.

The time diagrams of read and write cycles are shown in Figure 5.16; a read-modify-write cycle is used in VME data transfer protocol for indivisible operations in multiprocessor environments. The read-modify-write cycle is shown in Figure 5.17. The VME has no bus lock signals, but the bus can be locked by keeping AS* active.

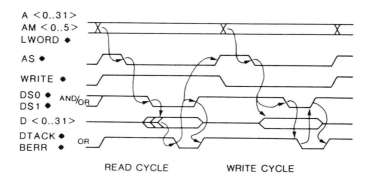

Fig. 5.16 – Read and write cycles

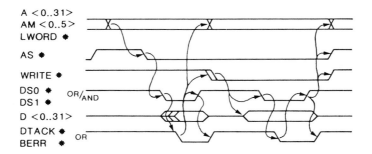

Fig. 5.17 – Read–modify–write cycle

5.3.4. Special Features

The VME bus provides a powerful interrupt handling structure. Seven interrupt request lines IRQ <7..0> with different priority levels are present on the bus. The slaves request interrupt to the specific master on one of these lines. Different masters with specific interrupt request lines are allowed on VME bus and many slaves can be connected to each request line. A daisy-chain is used to resolve multiple requests.

More than one master capable of handling interrupts can be placed on VME bus. When a master receives an interrupt request, it becomes the commander via a bus arbitration cycle, then performs an interrupt acknowledge cycle.

In the interrupt acknowledge cycle the commander, by means of an addressing-like operation, declares which interrupt request line is being honoured. The requesting slaves that identify their interrupt code will use the daisy-chain to find the highest priority requesting slave. The selected slave places its vector and acknowledges the transfer with the DTACK* line following the same asynchronous protocol of read cycles. Figure 5.18 shows a block diagram of the interrupt acknowledge structure, Figure 5.19 refers to the interrupt cycle timing.

In the VME bus standard two lines are reserved for a serial bus. It is defined as an optional feature, not essential to the system but useful as secondary control and data path. The bus hardware provides two lines, one for data and one for clock. Figure 5.20 shows the serial bus timing.

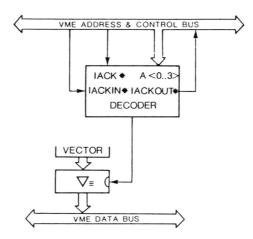

Fig. 5.18 – Block diagram of the interrupt acknowledge structure

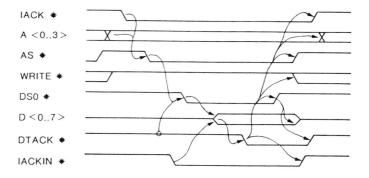

Fig. 5.19 – Interrupt acknowledge cycle

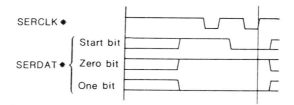

Fig. 5.20 – Serial bus timing

An optional feature on VME systems is the sequential memory access. The name used inside the VME specifications refers to the block transfer capability between the master and the slave during the data transfer cycle. Master and slave with this option have on board the circuitry to generate the sequential addresses to access data in the memory. Owing to the non-multiplexed nature of the VME bus, the block transfer capability shows no speed benefits with respect to the normal transfer.

The pinout of VME bus is shown in Figure 5.21.

	A	B	C
1	D00	BBSY*	D08
2	D01	BCLR*	D09
3	D02	ACFAIL*	D10
4	D03	BG0IN*	D11
5	D04	BG0OUT*	D12
6	D05	BG1IN*	D13
7	D06	BG1OUT*	D14
8	D07	BG2IN*	D15
9	GND	BG2OUT*	GND
10	SYSCLK	BG3IN*	SYSFAIL*
11	GND	BG3OUT*	BERR*
12	DS1*	BR0*	SYSRESET*
13	DS0*	BR1*	LWORD*
14	WRITE*	BR2*	AM5*
15	GND	BR3*	A23
16	DTACK*	AM0	A22
17	GND	AM1	A21
18	AS*	AM2	A20
19	GND	AM3	A19
20	IACK*	GND	A18
21	IACKIN*	(SERCLK)	A17
22	IACKOUT*	(SERDAT)	A16
23	AM4	GND	A15
24	A07	IRQ7*	A14
25	A06	IRQ6*	A13
26	A05	IRQ5*	A12
27	A04	IRQ4*	A11
28	A03	IRQ3*	A10
29	A02	IRQ2*	A09
30	A01	IRQ1*	A08
31	-12	+5 STDBY	+12
32	-5	+5	+5

Fig. 5.21 - Pinout of VME bus

5.4. THE 896 BACKPLANE BUS

5.4.1. History and Main Features

"P896" was the Project Authorization Request Number assigned in 1979 by the IEEE Standard Board to a committee set up with the purpose of defining a backplane for "future" microcomputer-based systems (the same project is known also as "Futurebus"). The design goal was a processor-independent backplane bus, able to support multiprocessor systems, with fully distributed control, a 32-bit transfer width, error detection capability, and a high transfer rate with asynchronous protocol.

Two backplane lines were reserved from the beginning to a serial transmission structure. This feature is now common to most multiprocessor buses, and was pioneered by P896.

Special care has been given to transmission line behaviour of backplane tracks. Reflections, crosstalk, wired-or glitches and the complete electrical behaviour have been considered in the design of the protocol and in the specification of transceivers. For this reason the constraints on the electrical characteristics of the backplane are more severe in P896 than in other buses, and the transceivers are non-standard devices, with parameters optimized for the bus environment.

A first draft (D4.1) of P896 was completed in 1981. Since then the specifications were deeply discussed by the committee, revised by external people, tested by prototype implementations, and compared with new industrial standards developed in the meantime. The last document (D6.2) |P89683|, completed in November 1983, contains complete specifications at the mechanical, electrical, and functional levels.

This Section is a summary description of P896; it gives a general overview of its design philosophy, with details of some specific operations. A detailed discussion of most relevant issues in P896 design is in |BORR84, TAUB84|.

5.4.2. Physical and Electrical Specifications

P896 uses standard Euro-size boards and 96-way DIN 41612 connectors, with females mounted on the backplane. All board sizes shown in Figure 5.22 are accepted. The mechanical specifications are derived from IEC rules for equipment using 19" racks |IEC249, IEC297, IEC603|. Each backplane mounts up to 21 connectors. The main 896 bus corresponds to connector A in Figure 5.22.

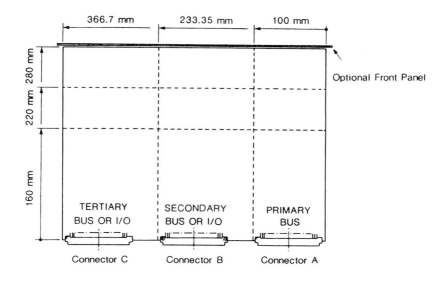

Fig. 5.22 - 896 board family

The only supply voltage distributed on the backplane is the main logic supply, + 5 V. Six pins on each connector are reserved for the + 5 V supply; another 6 are supply ground; 10 more pins are specified as separate signal grounds. All other voltages should be obtained by means of on-board DC-DC converters.

The unloaded characteristic impedance of each 896 track should be 50-60 ohms. Termination circuits at both ends, as shown in Figure 5.23, are required. The equivalent capacitance of each module must be 10 pF/line or less (this value cannot be met by standard TTL or CMOS transceivers). The propagation delay Tpd on a fully loaded backplane in these conditions is about 12 ns.

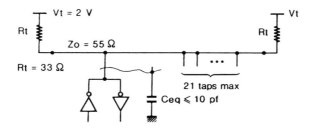

Fig. 5.23 - Equivalent circuit of a 896 track

The electrical levels of bus signals are shown in Figure 5.24 They
are specified as percentage of the main supply Vcc; this figure also
gives the values for Vcc = 5 V.

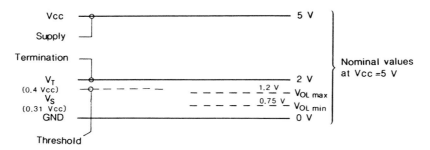

Fig. 5.24 – Electrical levels on 896 lines

Only open-collector drivers are used; their output current must
comply with the specifications of Figure 5.24 when driving a
terminated line. This requires about 50 mA static and 100 mA
dynamic, for Vo 1.2 V. Transition times must be 4–8 ns to limit
crosstalk.

On edge-active lines (that is on timing signals), the receiver
must include a noise-filtering integrator. A logic state change on
these lines is sensed only when the electrical level recrosses a
threshold for at least 2Tpd, as shown in Figure 5.25. This
guarantees that reflections and crosstalk do not cause false strobes.

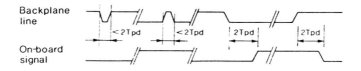

Fig. 5.25 – Input and output of integrating receivers

These electrical specifications for transceivers do not allow the use
of STTL or LSTTL devices; however, it is explicitly mentioned that
these devices can be tolerated on experimental boards.

The following procedure for live insertion is suggested:

- Apply Vcc to the board via a separate umbilical cable;
- disable the board;
- insert the board in the backplane;
- remove the umbilical cable;
- enable the board.

The reverse procedure (disable,   connect cable,   remove) applies for
live removal.

5.4.3. The Information Transfer Protocol

Each P896 information transfer consists of three sequential cycles:
arbitration, addressing, and data transfer.
     The arbitration uses a separate set of bus lines, and can be
carried on concurrently with addressing and data transfers.   These
last operations are multiplexed on a set of 32 common Address/Data
lines (A/D),   and use independent control signals.   The complete
sequence of bus operations is shown in Figure 5.26.

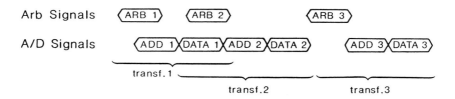

Fig. 5.26 - Examples of 896 bus operations, showing
sequencing of arbitration and A/D transfers

The arbitration structure of P896 is basically the same as 696/S100
|S10079| and Fastbus |FAST81|,   with some changes to make it fully
distributed, asynchronous, and fair. An upper limit for servicing of
bus requests is guaranteed in real-time applications because each
module can compete again for the bus only if all other pending
requests are satisfied.
     A block diagram of the complete 896 arbitration system is
shown in Figure 5.27a.   Each requesting unit issues a different
priority code on AN<6..0>*. AN6* distinguishes between priority and
fair units;   AN<5..1>* is a unique module identifier,   derived from
the geographic slot identifier of the backplane.   AN0* is a parity
flag which allows the checking of the correctness of arbitration
codes.   The contention process selects the highest priority unit with
the technique described in Section 4.2.3. The arbitration signal AC*
controls the   fairness mechanism.   Since   the arbitration   involves
N-partner information exchange (see Section 4.4.1),   the sequencing
of operations is synchronized by a three-wire handshake,   which
uses the signals AP*, AQ*, AR* as shown in Figure 5.27b.

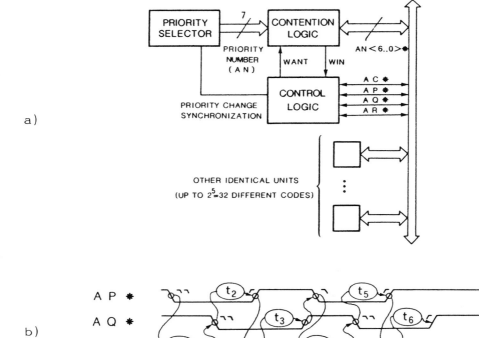

Fig. 5.27 - 896 arbitration
        a) Block diagram of the arbitration subsystem
        b) Arbitration handshake and sequence of operations

The priority arbitration procedure is as follows:

1) A master module requesting to become commander initiates the
    process by activating AP*.   Also all other units with pending
    requests activate AP* (state 1), and

2) enable their contention logic (state 2).   If the procedure starts
    during a data transfer, the current commander is not allowed to
    participate in the contention.   The contention is terminated when
    the slowest unit releases AP*.

3) Errors can be checked at this time (state 3).

4) If no unit detects errors AQ* is released and,   as soon as the
    bus is not busy (state 4),

5) the winner gets control of the bus (state 5),   becomes the
    commander and starts the address/data cycles.

6) The priority number of the winner is maintained on AN* lines
    (state 6),   and can be sampled by passive monitoring units to
    keep track of the sequencing of commanders.

The address/data protocol uses two-wire handshake for single-
responder operations,   and three-wire handshake for fully
asynchronous N-partner broadcast and broadcall.   The addressing
cycle can be followed by any number of data transfers,   either read
or write.   The signals used in the addressing cycle are:

    AS*   Address Strobe;
    AK*   Address Acknowledge;
    AI*   Inverted Address Acknowledge (only for N-partner
                transfers).

The information transmitted in the address cycle consists of 32
address bits, plus 5 mode bits with the following meaning:

    CM4*   read/write;
    CM3*   lock/unlock (for indivisible operations);
    CM2*   single/block transfer;
    CM1*   single responder/broadcast;
    CM0*   reserved.

The addressed slave responds with an acknowledge and a status
word ST<2..0>,   which specifies if the requested operation can be
successfully completed.
      After this answer the commander enters the data cycle.   The
data handshake signals are:

    DS*   Data Strobe;
    DK*   Data acKnowledge;
    DI*   Inverted Data acknowledge (only for N-partner
                operations).

The timing diagrams of combined address/data cycles are shown in
Figure 5.28.

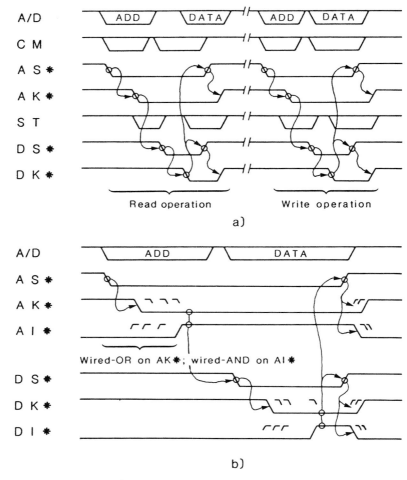

Fig. 5.28 – A/D trasfer cycles
         a) Single responder (AI* not used)
         b) Multiple responder write (3-wire handshake)

In the data cycles CM4* distinguishes read/write operations, while CM<3..0> are used as byte lane enable flags. They allow random operation on single bytes within the accessed 32-bit data word.

In block transfers both edges of DS*/DK* are used to enhance speed. Figure 5.29 shows an example of block transfer for an even number of data. Odd transfers are allowed with a different end-of-cycle sequence.

Error detection is achieved by means of 5 Error Detection bits (ED), which protect the A/D word and command lines, plus an Error Valid (EV) signal. ED bits are computed as byte parity.

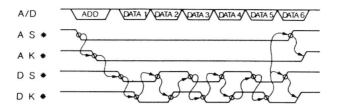

Fig. 5.29 – Block transfer (write)

A responder which detects an error notifies it to the commander using the status signal ST.

5.4.4. Special Features

The only global system control signal is the Reset (RE). Bus initialization, warm start, and cold start are distinguished by the duration of the reset pulse.

The 896 backplane has no provision for carrying interrupt signals. It is assumed that each board has a processor able to handle local I/O interfaces and, therefore, most of the bus traffic is either data block move or interprocessor messages. The latter can use, as low level interprocessor communication structure, a set of control registers which are accessed from the backplane as memory locations and which generate direct commands such as interrupts, reset, etc. towards their on-board processor.

Two lines are reserved to a serial bus. It is defined as an optional feature, not essential to system operation, provided mainly as an alternative control path in case of faults. At user choice, it can also be used as generic system communication utility.

The hardware level of the serial bus is capable of supporting a wide variety of protocols and message types. The two lines SB0 and SB1 carry separate clock and data, synchronized as shown in Figure 5.30. The clock is transmitted by a single unit, but redundant units can be used to guarantee system integrity.

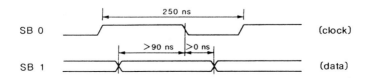

Fig. 5.30 – Clock / data synchronization on the serial line

The arbitration exploits the same contention mechanism of the parallel bus, implemented in a serial way on the data line SB1. Each transmitting unit compares its data with the line logic state to detect collision; a jammed unit retires within one clock period. This protocol is derived from I2C |MOEL80|, and guarantees a self-arbitrating transmission. Since all the SB1 line must become equipotential within a fraction of the clock period, the (length)X(data rate) product of the serial bus is limited.

Each message on the serial bus has the structure shown in Figure 5.31.

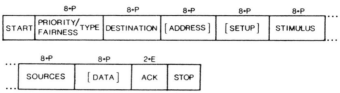

Fig. 5.31 - Message format

START and STOP are special symbols which bracket the whole message; P and E are parity bits. The parts within square brackets are inserted only in some messages.

The 896 specification defines different formats for messages, from system commands to user defined packets of unlimited length. A TYPE field also allows one to change the low-level protocol from the standard to a user-defined one. This makes a customized use of the serial bus possible, while keeping compatibility with the standard protocol for some functions.

5.4.5. Timing and Pinout

Owing to the fully asynchronous protocol, there is no timing constraint on modules, but only timing and electrical constraints on backplane and transceivers. These figures have already been discussed in Section 5.4.2.

The main limit for the bus speed is the delay introduced by integrating receivers. A timing diagram which puts in evidence this effect is shown in Figure 5.32a and 5.32b.

Signals are allocated to the 96-pin connector in such a way as to minimize errors caused by crosstalk. Edge-active lines are shielded by grounds or static lines, and critical pairs are kept at some distance. Each octal transceiver has a separate signal return path, to minimize ground noise. The complete pinout is given in Figure 5.33.

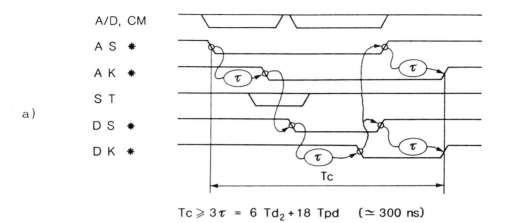

a)

$$Tc \geqslant 3\tau = 6\ Td_2 + 18\ Tpd \quad (\simeq 300\ ns)$$

b)

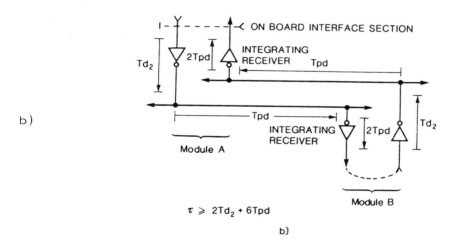

$$\tau \geqslant 2Td_2 + 6Tpd$$

b)

Fig. 5.32 – Influence of the handshake delay on the 896 timing
        a) Single write cycle (signals are shown at interface
           I, before bus drivers and after receivers
        b) Equivalent circuit for the evaluation of the delay

	A	B	C
1	GND	GND	GND
2	+5	+5	+5
3	AD0*	AD1*	AD2*
4	AD3*	GA0*	AD4*
5	AD5*	AD6*	AD7*
6	GND	ED0*	AD8*
7	AD9*	AD10*	GND
8	AD11*	AD12*	AD13*
9	AD14*	GA1*	AD15*
10	ED1*	AD16*	AD17*
11	GND	AD18*	AD19*
12	AD20*	AD21*	GND
13	AD22*	AD23*	ED2*
14	AD24*	GA2*	AD25*
15	AD26*	AD27*	AD28*
16	GND	AD29*	AD30*
17	AD31*	ED3*	GND
18	CM0*	CM1*	CM2*
19	CM3*	GA3*	CM4*
20	CP*	EV*	ST0*
21	GND	ST1*	ST2*
22	AS*	AK*	GND
23	AI*	DS*	DK*
24	DI*	GA4*	AP*
25	AQ*	AR*	AC*
26	GND	AN0*	AN1*
27	AN2*	AN3*	GND
28	AN4*	AN5*	AN6*
29	SB0*	RE*	SB1*
30	RFU0	RFU1	RFU2
31	+5	+5	+5
32	GND	GND	GND

Fig. 5.33 — Pinout of P896 bus

5.5. THE M3BUS BACKPLANE

5.5.1. History and Main Features

The design of M3BUS (Modular Multi Micro BUS) was an activity coordinated in the frame of a larger research project (Computer Science Program of the Italian National Research Council), and with other standardization efforts, especially with the IEEE-P896 committee. The goal was to define a processor-independent system bus for high performance multiprocessor machines based on 16-bit micros like Z8000, MC68000, IAPX186/286, and others.

The first M3 specification was released in July 1980 |M3BU81|, and this first version has been used in the MUMICRO and MODIAC projects of the above-mentioned Computer Science Research Program. After the first experimental implementations M3 is now frozen to the specifications given in this section. Owing to cooperation with P896, some ideas are shared by the two designs; in particular M3BUS has many common features with the P896 specification D4.1, developed in 1980/81. Since then, P896 has evolved into a 32-bit bus, while M3 stayed with 16-bit machines. M3 is now used in Italy by companies which develop systems for industrial automation and control.

M3BUS is specifically oriented to multiprocessor systems; it consists of two independent buses for information transfers, a parallel and a serial one, both residing on the same backplane. The parallel bus is the main path for high speed data transfers. It is used for arbitration, addressing, data transfer, to carry interrupt requests and to perform other special cycles provided for interprocessor communication. M3 uses 24-bit addresses and 16-bit data; a 32-bit version is being defined |DELC84| and is not described here. The serial bus is used for communications among intelligent units; it mainly carries system messages. Both buses have a hardware error detection capability to allow for the organization of fault-tolerant systems.

The complete specifications of M3BUS and a system design guide are in |CICO83|.

5.5.2 Physical and Electrical Specifications

The size of the cards used in M3BUS systems follows the IEC 297 specification shown in Figure 5.34. The preferred size is double Eurocard, 220 mm in length. The connectors used are DIN 41612 version C, with the male mounted on the card. The bus can operate with only 64-pins on rows a and c. The central row b has been

reserved for additional ground and supply pins, optional signals
and future extensions.

The backplane is a printed circuit board carrying 96 bused
tracks with a maximal length of 500 mm.

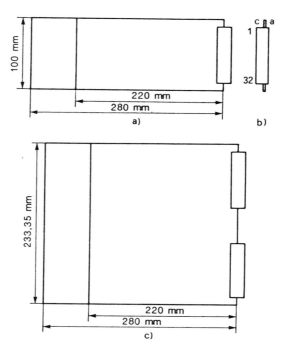

Fig. 5.34 - M3BUS boards (component side views)
                    a) single eurocard
                    b) DIN connector (front view)
                    c) double eurocard

The supply voltages distributed in M3BUS are:

+5 V   Main logic supply (5 pins);
+15V   Interface supply (1 pin);
−15V   Interface supply (1 pin);
+5 V   Backup for logic (1 pin).

The logic backup is intended to provide a continuous supply for
circuits which cannot be switched off, such as permanent memories,
calendar-clocks, etc. The maximum current per pin in M3 connectors
is 1 A. The ground pins act as a return path for both supply and
signals; their number must be higher than supply pins, and 12
ground pins are provided on the 96-pin connector.

Unloaded   bus   lines   (except   power   supplies)   must   have   a
characteristic impedance between 50 and 150 ohms.   The lines are
terminated with one of the circuits shown in Figure 5.35.

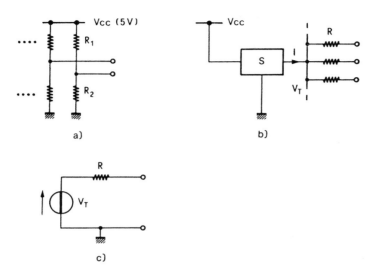

a)   b)

c)

Fig. 5.35 – Termination circuitry
            a) Passive network
            b) Active network.
                With 3-state drivers the current I can be
                negative;   the regulator S must both source
                and sink current.
            c) Equivalent circuit
                $R=400\ \Omega$ if terminated at both ends
                $R=200\ \Omega$ if terminated at one end
                $V_T=3.3$ V

M3BUS   uses   three-state   drivers   for   address/data   and   for   some
control lines;   open collector drivers are used for the lines which
support the wired-or function on the bus.   The voltage levels on the
backplane are summarized in Figure 5.36.
        Bus drivers must have leakage currents no greater than $20\mu A$
(3-S) or 100 µA (OC) at high level,   and -400 µA at low level.   At
the voltage levels specified in Figure 5.36,   the low level output
current must be 48 mA at least;   the high level current (specified
for 3-S only) must be at least -3 mA.
        The load of a receiver on a backplane line must not be
greater than 20 µA at 2.4 V and -400 µA at .5 V.   Each board
plugged into M3BUS should have a maximum of 2 receivers and one

driver per line. Two 3-S drivers are tolerated to simplify the
design. With the figures mentioned above, up to 27 boards can be
tied to the same bus segment.

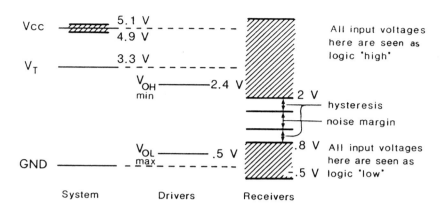

Fig. 5.36 – Voltage levels in M3BUS

5.5.3. System Organization and Control

M3BUS is multiplexed and all transactions take at least three
cycles, one for arbitration, one to transfer the address and one to
transfer the data, as shown in Figure 5.37.

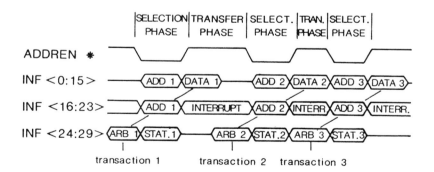

Fig. 5.37 – Multiplexing and pipelining in M3BUS

The parallel information bus consists of 30 general purpose
information lines (INF<0:29>), plus 4 parity lines (PAR<0:3>) and
other control lines. The bus signal ADDREN* indicates when the bus
is in the selection phase or in the transfer phase. Inside each
phase, the other signals control the action sequences for

addressing,   arbitration,   data transfer and interrupt requests. The
details of these operations are discussed in sections 5.5.4,...8.

Some system control signals carry   general   purpose information
usable by any module,   independently of the information transfer
protocol. They are the following ones:

RESET*
This signal forces the system into a known state.   The timing
specifications for RESET* are shown in Figure 5.38.

SCK (System ClocK)
General purpose timing signal,   provided by one module only.   The
arbitration logic must use SCK for synchronization;   all other bus
signals can be asynchronous with the system clock,   however it is
possible to use this clock for sequential logic in any module.
Currently the frequency of SCK in M3 systems is 4 MHz.

PROCDW* (PROCessor DoWn)
This signal becomes active after a fatal error or a local power
failure   in a processor board.   It initiates a system start up or
reconfiguration.

PWFAIL* (PoWer FAILure)
This signal is activated by the power supply subsystem when the
AC main leaves its nominal values to signal a possible impending
power failure.

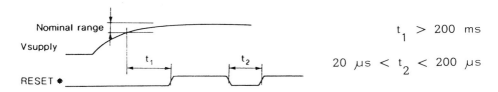

$$t_1 > 200 \text{ ms}$$

$$20 \ \mu s < t_2 < 200 \ \mu s$$

Fig. 5.38 – Timing of RESET* signal

5.5.4. The Arbitration Protocol

In M3BUS the arbitration is performed by a distributed arbiter with
coded priority |TAUB76|.   The INF<24:27> bus lines are used to
carry arbitration priorities during the transfer phase,   therefore up
to 16 priority codes can be handled.   This value is a trade-off

between the capability of having many different priority levels and the speed of the arbitration process. The control signals used by the arbitration logic are:

SCK; defined in Section 5.5.3.

BBUSY* (Bus BUSY)
This signal is active when a master holds the bus for addressing or data transfer cycles.

BREQ* (Bus REQuest)
This signal is active when an access request is pending.

The behaviour of a master module which wants to become commander is shown in Figure 5.39.

CONDITIONS	ACTIONS
1) Request pending	BREQ*=0
2) Transfer phase (ADDREN*=1) System clock	Put priority code on INF 24:27 and arbitrate
3) System clock WIN active Transfer phase (ADDREN*=1) Bus free (BBUSY*=1)	Signal bus busy (BBUSY*=1) Remove priority code BREQ*=1 Get the bus Start selection cycle (ADDREN*=0) Perform the bus transaction Signal bus free (BBUSY*=1)

Fig. 5.39 – Arbiter behaviour

It starts by requesting the arbitration circuitry to begin an
arbitration cycle. As soon as the system is in a transfer phase, an
arbitration cycle can start on the rising edge of SCK. Every master
requesting the bus puts its own code on the priority lines, and the
self-selection network carries out the arbitration. At the following
rising edge of SCK, the WIN output of the active priority networks
is sampled in each master. Since the self-selection process has been
completed, only one WIN is active, and this master is allowed to
become the commander for the next transaction. The winning master
waits until the bus is free by looking at the bus busy (BBUSY*)
line, then it occupies the bus by activating BBUSY*, and begins the
selection phase. At the end of the bus transaction, the commander
releases the bus control and deactivates BBUSY*.

The masters which have lost the arbitration wait for the next
arbitration cycle, which will start in the following transfer phase.
The bus signals of an arbitration cycle are shown in Figure 5.40.

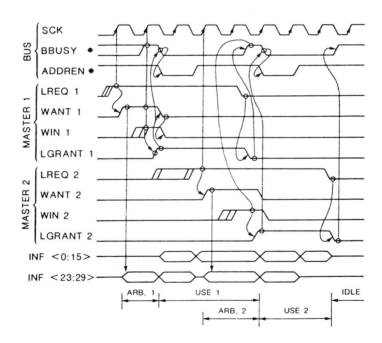

Fig. 5.40 – Example of arbitration cycles

This protocol can be upgraded by using the BREQ* signal to
indicate if any module is queueing for the bus; it allows
implementation of a "fair" arbitration as explained in Section 4.2.3.
Fairness avoids the problem of modules with low priority codes

waiting for an unbounded time because masters with higher priority
monopolize the bus with many consecutive requests. To get fair
priority, each module waiting for the bus activates the BREQ*
signal: a module which has already had access to the bus cannot
make another arbitration cycle as long as the BREQ* stays active,
because this condition means that another module is still queueing
for the bus. An M3 fair arbiter is described in |CIVE83|.

5.5.5.   The Addressing Protocol

The addressing operation allows the commander to identify one or
more partners for the following data transfer. It is performed in
the selection phase of the M3BUS transaction. The basic addressing
cycle in M3BUS is synchronous, and all the timing is guaranteed by
the master; the slaves must be fast enough to catch the address
information. The bus signals used in the basic addressing phase
are:

ADDREN* (ADDress ENable)
This signal is active when the information lines can be used for
address (INF<0:23>) and status (INF<24:29>).

CYCLE*
The falling edge of this signal indicates that the information
carried by the INF lines during the addressing cycle is valid. The
rising edge is used in the transfer phase to signal the end of the
bus transaction.

The action sequence and the action mapping into bus signals of the
addressing cycle are given in Figure 5.41a. In order to address the
responder, the commander signals the beginning of the selection
phase by activating the ADDREN* line, then it puts the address,
the status, and possibly the 4 parity signals on the bus. After the
propagation, settling and set-up times plus a margin to allow for
possible supervisor intervention, the commander validates address
and status information with the falling edge of the CYCLE* signal.
After this step it waits for the minimum guarateed hold time and
then starts the data phase by deactivating ADDREN*. The timing
diagram of the addressing cycle is shown in Figure 5.41b. In I/O
transactions only the lower 16 address bits (INF<0:15>) are used.
        The status information issued by the master on INF<24:28>
 indicates the type of operation to be performed, according to
Figure 5.42.

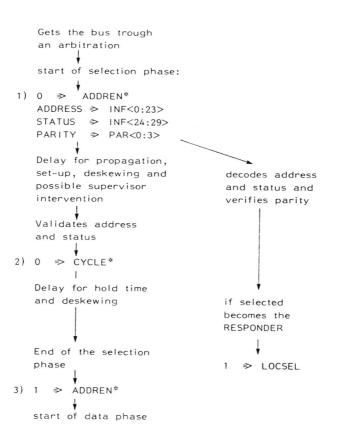

COMMANDER   RESPONDER

Gets the bus trough
an arbitration

start of selection phase:

1) 0 ⇒ ADDREN*
    ADDRESS ⇒ INF<0:23>
    STATUS ⇒ INF<24:29>
    PARITY ⇒ PAR<0:3>

    Delay for propagation,
    set-up, deskewing and   decodes address
    possible supervisor   and status and
    intervention   verifies parity

a) Validates address
    and status

2) 0 ⇒ CYCLE*

    Delay for hold time
    and deskewing   if selected
                                  becomes the
                                  RESPONDER

    End of the selection
    phase   1 ⇒ LOCSEL

3) 1 ⇒ ADDREN*

    start of data phase

b)

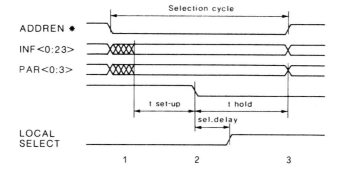

Fig. 5.41 – Addressing cycle
a) Action sequence
b) Timing diagram

INF29 27 26 25 24

0	0	0	0	0	Reserved
0	0	0	0	1	CPU memory access, operand, indivisible
0	0	0	1	0	" " " stack "
0	0	0	1	1	Reserved
0	0	1	0	0	DMA 0 " " "
0	0	1	0	1	" 1 " " "
0	0	1	1	0	" 2 " " "
0	0	1	1	1	Block " " "
1	0	0	0	0	CPU " " fetch
1	0	0	0	1	" " " operand
1	0	0	1	0	" " " stack
1	0	0	1	1	Reserved
1	0	1	0	0	DMA 0 " "
1	0	1	0	1	" 1 " "
1	0	1	1	0	" 2 " "
1	0	1	1	1	Block " "
1	1	0	0	0	Interrupt acknowledge, Vectored Interrupt
1	1	0	0	1	" " Non Vectored Interrupt
1	1	0	1	0	" " Non Maskable Interrupt
1	1	0	1	1	Segment Trap
1	1	1	0	0	Normal Input Output
1	1	1	0	1	Special Input Output
1	1	1	1	0	Reserved
1	1	1	1	1	Bus Idle

Fig. 5.42 - Status encoding

INF29 at 0 indicates an indivisible operation, INF28 specifies if the master is in normal (1) or system (0) state, INF27 distinguishes memory (0) or non-memory (1) operations.

5.5.6. The Data Transfer Protocol

The data cycle begins when the current master releases the ADDREN* line and terminates when it releases the CYCLE* bus signal. The bus signals used in M3BUS for the data transfer are the following ones:

LODAVAL* (LOw DAta VALid)
When active signals that INF<0:7> lines carry valid data.   It is
used by responders as strobe in write operations and as output
enable in read operations.

HIDAVAL* (HIgh DAta VALid)
Same as LODAVAL*, for INF<8:15>.

WRITE*
Active when the current bus operation is a data transfer from the
commander to the responder(s) (write cycle).

TRACK* (TRansfer ACKnowledge)
Handshake signal of data transfer, activated by the responder(s).

BRACK (BRoadcast ACKnowledge)
Second handshake line, active high, which implements a wired AND,
allowing the delay of the end of cycle up to the last acknowledge
from a selected slave.

The data transfer cycle uses an asynchronous protocol.   The write
operation is signalled by WRITE* line active (low level) and follows
these rules:

1) As soon as the responder is selected,   it deactivates BRACK to
    indicate that it is waiting for data.

2) The commander puts the information on INF<0:15> and validates it
    by activating either HIDAVAL* or LODAVAL* as requested by the
    data format (word, high byte, low byte).

3) When the responder has accepted the data,   it activates both
    TRACK* and   BRACK.

4) The commander terminates the cycle by deactivating the data
    valid lines,   takes the data away from the bus,   and closes the
    cycle by rising the CYCLE* signal.

The action sequence and the timing diagram of a write transfer are
shown in Figure 5.43.

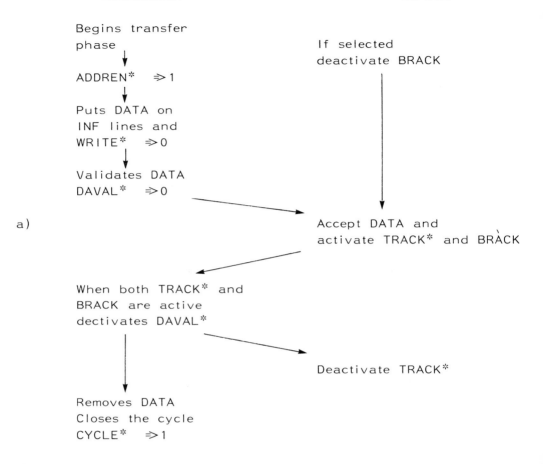

COMMANDER   SLAVES

Begins transfer
phase   If selected
                                  deactivate BRACK
ADDREN*   ⇒1

Puts DATA on
INF lines and
WRITE*   ⇒0

Validates DATA
DAVAL*   ⇒0

a)
                                  Accept DATA and
                                  activate TRACK* and BRACK

When both TRACK* and
BRACK are active
dectivates DAVAL*

                                  Deactivate TRACK*

Removes DATA
Closes the cycle
CYCLE*   ⇒1

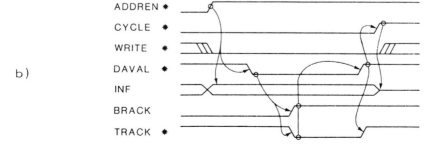

ADDREN *
CYCLE *
WRITE *
b)
DAVAL *
INF
BRACK
TRACK *

Fig. 5.43 – Write operation
              a) Action sequence
              b) Timing diagram

The read operation is indicated by WRITE* inactive (high level) and follows these rules.

1) As soon as the responder is selected it deactivates BRACK to indicate that the data on bus are not yet valid.

2) The commander requests the information (word or byte) by activating as necessary HIDAVAL* and LODAVAL* signals.

3) The responder puts the data on the bus and validates them by activating both TRACK* and BRACK.

4) The commander accepts the data and terminates the operation by deactivating the data valid lines, then it closes the cycle by rising the CYCLE* signal.

The action sequence and the bus signals of a read transfer are shown respectively in Figure 5.44a and 5.44b.

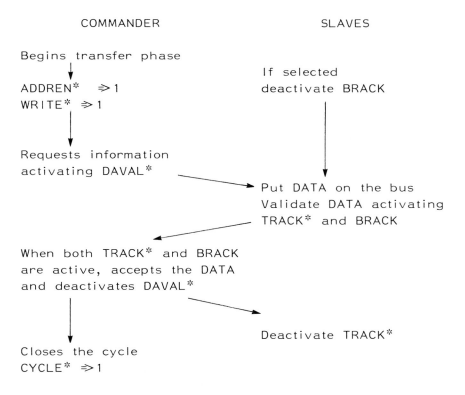

COMMANDER   SLAVES

Begins transfer phase
                                    If selected
ADDREN*   ⇒ 1   deactivate BRACK
WRITE*   ⇒ 1

Requests information
activating DAVAL*
                                    Put DATA on the bus
                                    Validate DATA activating
                                    TRACK* and BRACK

When both TRACK* and BRACK
are active, accepts the DATA
and deactivates DAVAL*

                                    Deactivate TRACK*
Closes the cycle
CYCLE* ⇒ 1

Fig. 5.44a – Action sequence of a read operation

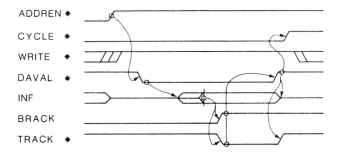

Fig. 5.44b - Timing diagram of a read operation

The M3 protocol also allows the use of special cycles such as broadcast transfers, indivisible read-modify-write, read-after-write cycles, and block transfers.

In broadcast operations, the data transfer is controlled by the wired-AND and by the wired-OR respectively on the lines BRACK and TRACK*. The timing diagram of a broadcast write operation is shown in Figure 5.45.

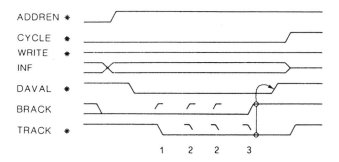

Fig. 5.45 - Timing diagram of a broadcast write operation

1) The fastest destination accepts the information: TRACK* goes active, BRACK stays indicative.

2) Medium speed destinations accpet the information: TRACK* and BRACK lines are not affected.

3) The slowest desination accepts the information and activates BRACK: the commander detects than the information has been accepted by every partner and closes the cycle.

A block transfer cycle is composed of an arbitration phase, an addressing phase, and a data transfer phase in which more than one read or write operation is performed, as shown in Figure 5.46. It allows for faster access of consecutive memory locations than the usual transactions, because the addressing cycle is not repeated.

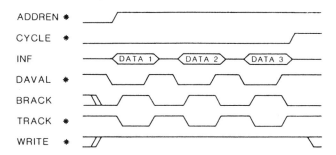

Fig. 5.46 – Block transfer operation

Information transmitted on the bus is protected by using odd byte parity. The parity signals are generated and checked only for the sets of INF lines which actually carry information in the current phase. Parity is optional, and boards with or without parity can be mixed. The boards which generate parity activate the parity enable signal PAREN*, thus enabling the parity to be checked by the modules which receive the information.

5.5.7. Interrupt and Inter-Processor Communication

In M3BUS there are two kinds of interrupt requests coming from peripherals and directed to one or more masters; they correspond, respectively, to non-maskable and maskable interrupts. Each request is directed by hardware to a single master only, but it is also possible for a processor to send an interrupt to another processor under software control.
    The non-maskable interrupt request has a dedicated bus line (NMI*). After a non-maskable interrupt request, the master initiates an NMI acknowledge sequence, composed of arbitration, selection, and a data transfer phase. In the selection phase, the status information carried by <INF 24:29> indicates NMI acknowledge. The following data phase transfers a 16-bit vector from the peripheralto the master as shown in Figure 5.47.

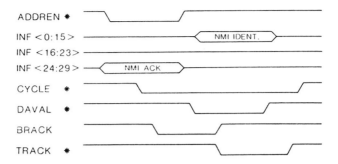

Fig. 5.47 – Non maskable interrupt acknowledge

The maskable interrupt requests are activated by the peripheral needing service by lowering one of the <INF 16:23> lines during the transfer phase (ADDREN*=1). Eight different priority levels are allowed. When an interrupt request is sensed, and the master is enabled to serve it, an interrupt acknowledge cycle is entered. The block diagram of the interrupt structure, and the complete interrupt acknowledge operation are shown in Figure 5.48. In the transfer phase the INF<8:15> lines carry information from the master to the slaves, and they indicate which level of interruption the master is acknowledging. The INF<0:7> lines carry information in the other direction, signalling which slaves of the selected level are requesting interrupt.

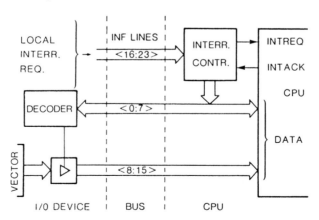

Fig. 5.48a – Block diagram of the interrupt structure

M3BUS provides a special cycle for events or command transmission. This cycle allows the transfer of a 30-bit vector on INF<0:29> from a commander to every other unit provided with a special

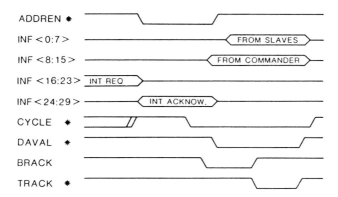

Fig. 5.48b – Timing diagram of the interrupt acknowledge

processor control register (PCR). The vector contains the address of the destination master, some predefined fields for direct commands, and a general purpose information field. The inter-processor interrupt cycle is composed of a regular arbitration cycle, where the sending master gains the bus control, a special addressing cycle, in which the commander writes the vector into one or more PCR, and a dummy data cycle, in which the commander does not take any action, but an arbitration for the next transaction can start. The special selection phase is a parallel synchronous transfer similar to an addressing cycle, but, instead of CYCLE*, the commander activates the PROCINT* (PROCessor INTerrupt) strobe, as shown in Figure 5.49.

Fig. 5.49 – Interprocessor interrupt cycle

5.5.8. Supervisor Protocol

The M3 parallel bus protocol exploits the Enable/Disable technique described in Section 4.5 to allow the insertion of special hardware modules called SUPERVISORs. A supervisor can slow the operations and replace the information (address and data) both in the selection and in the transfer phases. Supervisors use two dedicated lines on bus backplane: SUPervisor ON (SUPON*), and INHIBit (INHIB*).

In the addressing cycle, SUPON* acts as not-valid signal. When it
is active,   the slaves do not accept the address,   and the
commander stays in the selection phase.   When the SUPON* becomes
inactive again, the slaves accept the address on the bus, and the
commander,   after the address hold time,   terminates the   selection
phase.   An addressing cycle slowed down by a supervisor is shown
in Figure 5.50.

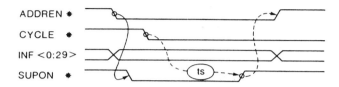

Fig. 5.50 – Addressing cycle slowed by a supervisor

For address replacement both SUPON* and INHIB* lines are used.
INHIB*,   when active in the selection phase,   disables the bus
drivers of the current commander.   In a cycle with address
replacement,   the supervisor inhibits the slaves from accepting the
address by activating SUPON* at the beginning of the cycle.   Then
the supervisor reads the address issued by the current commander
and validated by the falling edge of CYCLE*,   disables the master
address buffers using INHIB* line,   issues the new address,   and
signals that the address is valid by deactivating the SUPON* line.
The bus signals and action sequence of address replacement are
shown in Figure 5.51.

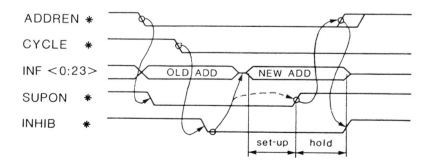

Fig. 5.51a – Timing diagram of address replacement cycle

COMMANDER   SUPERVISOR   SLAVES

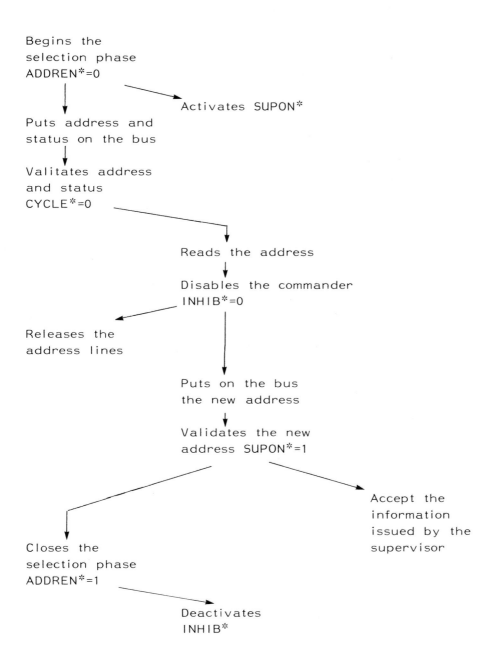

Begins the
selection phase
ADDREN*=0

Activates SUPON*

Puts address and
status on the bus

Valitates address
and status
CYCLE*=0

Reads the address

Disables the commander
INHIB*=0

Releases the
address lines

Puts on the bus
the new address

Validates the new
address SUPON*=1

Accept the
information
issued by the
supervisor

Closes the
selection phase
ADDREN*=1

Deactivates
INHIB*

Fig. 5.51b – Action sequence for address replacement

The write protection is obtained with the bus line SUPON*, which, if active, prevents the slaves from accepting data. A write protect operation is shown in Figure 5.52.

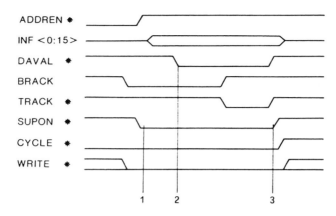

Fig. 5.52 – Timing diagram of a write protect operation
1) The supervisor disables data storage at slaves;
2) Data is not stored because SUPON* is active;
3) The cycle is closed when SUPON* becomes not active

The data replacement is obtained again by using the INHIB* line. In the data phase, this line disables the data buffers connected to lines INF<0:15> and the parity signals PAR<0:1> The bus signals of a data replacement in a read operation are shown in Figure 5.53. Other operations allowed by the M3 supervisor protocol are described in |DELC82|.

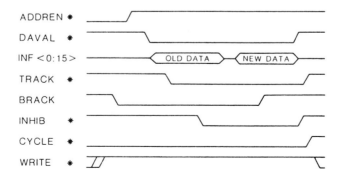

Fig. 5.53 – Timing diagram of a read operation with data replacement.

5.5.9. The Serial Bus

The serial bus connects all boards plugged into the M3 backplane. It uses two lines on the same connector of the parallel bus. The serial lines are driven by open collector devices and have the same electrical characteristics as the other lines. The serial bus is a multi-master multi-slave communication subsystem with no centralized elements. An arbitration mechanism ensures that only one master at a time can control the   bus.

    The   serial   bus   uses   a   modified   Inter   Integrated   Circuit protocol (I2C, |MOEL80, DERA84|). A device which performs most of the functions assigned to the M3 serial bus is the Philips-Signetic MAB8400 microprocessor.   The serial bus uses two lines: a self-synchronizing serial clock SERCK, and a data line SERDAT.

    The signal SERCK is generated by all the modules connected to the   serial   bus   with   the   technique   shown   in   Figure   5.54.   Each interface   has two timers,   one to define the duration of HIGH state (Th) and one for the duration of LOW state (Tl).   When the SERCK line goes from HIGH to LOW,   on each device the LOW timers are started   and   the   SERCK   line   stays   LOW   for   the   duration   of   the longest Tl,   than returns   HIGH. This transition triggers all the Th timers.   When the shortest Th is terminated,   the SERCK line is driven in the LOW state: this resets all the Th timers and starts all the Tl timers, so completing the cycle. Due to the open collector technique,   on   the   SERCK   line   the   HIGH   state   is   defined   by   the shortest Th of all modules,   and the LOW state is defined by the longest among Tl of the modules.

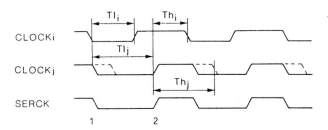

Fig. 5.54 – Self adaptive clock
        1) All Tl timers are started and Th timers reset;
        2) All Th timers are started

The data transmitted on SERDAT line are synchronized by the clock line.   Four different symbols which represent respectively data bit zero,   data bit one,   start sign,   and stop sign can be transmitted, as shown in Figure 5.55.

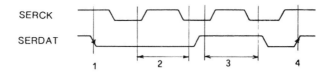

Fig. 5.55 – Example of a simple message on the serial line
1) Start sign;
2) Data cell (0);
3) Data cell (1);
4) Stop sign.

In 2 and 4 data can change,
in 3 and 5 data are valid.

Each message begins with a start sign,   contains a fixed number of
data bits   and terminates with a stop sign,   as shown in Figure
5.56.

8+1	8+1	8+1	8+1	8+1	2	
START	MASTER IDENTIFIER	DESTINATION IDENTIFIER	DATA	DATA	DATA	ACK NACK	STOP

Fig. 5.56 – Message format

A master can assume the control of the serial bus only if the bus
is free (i.e.   a clock cycle after a stop sign).   When two or more
masters see the bus free and begin to send at the same time their
identifiers,   an arbitration on a bit per bit basis is performed.
When a master transmits a LOW level while another master transmits
a HIGH level,   the result on the open collector bus line is a LOW
level.   The master transmitting the HIGH level notices the difference
between   the   internal   data   and   the   bus   level   and   retires
immediately;   the master which is not jammed maintains the control
of the bus.

The second byte of a message specifies the destination of the
message itself.   After this byte,   there are three free bytes for user
information.   At the end of the data transmission, two bits are sent
from the selected slave(s) to the master.   The first bit is a not
acknowledge bit (NACK),   and the second one is the acknowledge
(ACK);   both are active low.   This allows for the distinction of
correct operations (NACK*=1,   ACK*=0) from errors (0,1),   no slave
selected (1,1), and missed broadcast (0,0).

The start sign is generated by the master. The two first bytes

are generated by the master and go from the master to the slaves.
The three last bytes are generated by the master in write
operations and by the slave in read operations. The two
acknowledge bits are always generated by the selected slave(s).
The stop sign must be generated by the master. In byte
transmission, the most significant bit is transmitted first.

5.5.10. Timing and Pinout

The timing constraints of M3BUS are defined in Figure 5.57. They
have been derived for boards with two receivers/transmitters
connected to each line and under the following assumptions;

Propagation time on a 50 cm bus, driven and terminated as
defined in Section 5.5.2   20 ns

Settling time on the bus in the same conditions   50 ns

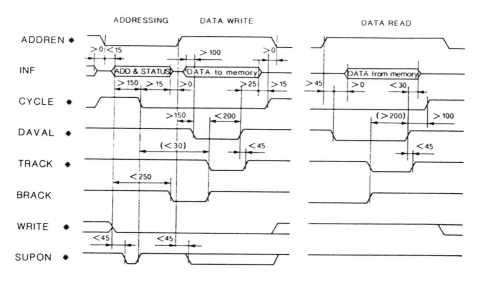

Fig. 5.57 − Timing of M3BUS

With these assumptions, the maximum delay between the emission of
a signal on the bus and the time when every module reads the
information correctly is   TB =70   ns

Delay for each logic level and driver   TL = 15 ns

Buffer disable delay   TD = 30 ns

Skew introduced by one level of logic   TK = 15 ns

Register set-up time   TS = 20 ns

Register hold time   TH = 10 ns

The complete pinout of M3BUS is given in Figure 5.58.

	A	B	C
1	GND	GND	GND
2	+5	+5	+5
3	INF 0	res	INF 1
4	INF 2	res	INF 3
5	INF 4	res	INF 5
6	INF 6	res	INF 7
7	INF 8	res	INF 9
8	INF10	res	INF11
9	INF12	res	INF13
10	INF14	res	INF15
11	INF16	res	INF17
12	INF18	res	INF19
13	INF20	res	INF21
14	INF22	res	INF23
15	INF24	res	INF25
16	INF26	res	INF27
17	INF28	res	INF29
18	PAREN*	res	DAER*
19	WRITE*	GND	CYCLE*
20	LODAVAL*	GND	HIDAVAL*
21	ADDREN*	GND	IRACK*
22	PAR 0	res	BRACK
23	PAR 1	res	PAR 3
24	PAR 2	res	SERCK*
25	SUPON*	res	INHIB*
26	BREQ*	GND	BBSY*
27	PWFAIL*	GND	SERDAT*
28	NMI*	GND	RESET*
29	PROCDW*	res	-15
30	+15	PROCINT*	SCK
31	+5	+5	+5 backup
32	GND	GND	GND

Fig. 5.58 - Pinout of M3BUS

5.6. REFERENCES

|BORR84| Borrill, P., and Theus, J., "An Advanced
             Communication Protocol for the Proposed IEEE 896
             Futurebus", IEEE Micro, August 1984.

|CICO83| Civera, P. et al, "Multiprocessors: M3BUS systems and
             TOMP architectures" PFI–MUMICRO report, September
             1983.

|CIVE83| Civera, P. et al, "An integrated self–selection
             arbiter", EUROMICRO 83 Proc., Madrid, October 1983.

|DELC82| Del Corso, D., Maddaleno, F., "Extension of bus
             protocols: a technique for modular upgrade of
             processing systems", EUROMICRO 82 Proc., Haifa,
             September 1982.

|DELC84| Del Corso, D., "Extension of M3BUS for 32–bit
             processors", PFI–MUMICRO Report, September 1984.

|DERA84| Del Corso, D., et al, "An integrated controller for
             modified inter integrated circuit protocol", Politecnico
             di Torino, DE, Internal Report, June 1984.

|FAST81| U.S. NIM Committee, "FASTBUS tentative specification",
             August 1981.

|IEC249| IEC publication 249–2, Printed Boards.

|IEC297| IEC publication 297–3, Mechanical structures, racks,
             and plug–in units; 297–3 Supplement A: Mounting of
             connectors.

|IEC603| IEC publication 603–2, Two–part connectors for printed
             boards.

|MOEL80| Moelands, A.P.M., "Serial I/O with the MAB8400 series
             microcomputers", Philips Electronic Components and
             Applications, Vol 3, No. 1, November 1980.

|MULT79| "Intel MULTIBUS specification", 1979.

|M3BU81| Del Corso, D., Duchi, G., "M3BUS: System specification for high performance multiprocessor machines", BIAS 81 Proc., Milan, October 1981.

|P89683| IEEE P896 Committee, "Futurebus – Specifications for advanced microcomputer backplane buses: P896-1, Draft 6.2", November 1983.

|S10079| Elmquist, K.A., et al., "Standard Specification for S-100 Bus Interface Devices", IEEE Computer, July 1979.

|TAUB76| Taub, D.M., "Contention resolving circuits for computer interrupt systems", Proc. IEE, num. 9, September 1976.

|TAUB84| Taub, D.M., "Arbitration and control acquisition in the proposed IEEE 896 Futurebus", IEEE Micro, August 1984.

|VMEB81| Mostek Corp., Motorola Inc., Signetics/Philips, "VME bus specification manual", 1981.

CHAPTER 6

HARDWARE MODULES FOR MULTIPROCESSOR SYSTEMS

D.Del Corso, M.Zamboni
Dipartimento di Elettronica
Politecnico di Torino
Torino, ITALY

ABSTRACT.   This chapter describes the basic features of hardware
modules for multiprocessor systems.   The physical organization of
those systems is examined,   and an analysis of module structure is
performed.   The approach here followed is not oriented to specific
buses but gives general design information limited to the block
diagram level.

6.1. INTRODUCTION

The design of a processing system according to a well defined
standard puts some constraints to the logical structure and the
implementation of modules such as boards,   backplane,   and other
sub-units.   These constraints apply both to the logical structure of
modules and to their implementation.   For instance,   the mechanical
specifications which define the size of boards also has an impact on
the number of packages that can be placed on them,   i.e.   on the
complexity of the module. The electrical specification, combined with
the characteristics of interface devices,   limits the number of units
connected to a single bus segment,   etc.   The information exchange
protocol defines the design of interface circuits,   and the space
required by these circuits impacts the amount of core devices
(processor,   memory,   I/O interfaces),   that can be placed on each
board.
        Within these constraints it is possible to identify some basic
structures and general design rules which must be followed to
obtain properly designed and reliable hardware.
In this chapter we shall analyze the basic organization of various
types of modules: master, slaves, mixed and special units.

G. Conte and D. Del Corso (eds.), Multi-Microprocessor Systems for Real-Time Applications, 225–278
© 1985 by D. Reidel Publishing Company.

As far as possible, the structures here defined are related only to the function performed, rather than to a specific bus standard. For this purpose, most of the examples are limited to the block-diagram level. Gate-level implementation depends strongly on the technology used (SSI/MSI, PALs, gate arrays, VLSI); we shall go down to this design level only for those aspects which depend on or influence the protocol itself. However, some more detailed examples for the buses presented in Chapter 5 are given.

In a complex multiprocessor system, some functions are supported by modules which cannot be classified as simple masters or slaves. The structure and design criteria for some of these units are outlined in Section 6.6.

6.2. SYSTEM DESIGN

6.2.1. Physical Organization of Multiprocessor Systems

In most of the multiple-bus multiprocessor systems described in Chapter 1, modules must be connected to at least two different communication structures. If both are parallel buses, with a lot of lines, the usual organization of modular processing systems based on a single backplane is no longer valid, and new mechanical setups must be used. The possibility to connect two different buses to a unique module is satisfactory in most cases; for all the standards described in Chapter 5 the problem has been considered and solutions have been proposed or are already in use.

To define the suitable organization we start from the fact that in most multiprocessor systems buses are organized hierarchically;

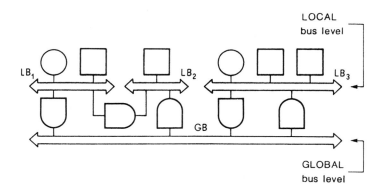

Fig. 6.1 - Bus hierarcy in a multiprocessor system

as shown in Figure 6.1, we can therefore identify one bus which carries the global communications, and a set of buses which interconnect a processor with some memory and interfaces. We shall call the former Global Bus (GB), and the latter Local Buses (LBs).

The GB must eventually go to all boards, while each LB is connected to some modules only. The usual solution is to group together the modules associated with the same LB, either on the same board or on adjacent boards, to make a "physically local" bus.

A first possibility is to use local buses confined within a board. Being designed for a fixed configuration, an on-board bus can be optimized for the set of devices used; synchronous protocols without buffering are the best choice in most cases. The drawback is that the number of units connected to an on-board bus is limited, and without possibility of expansion. Some reconfiguration can be obtained with the piggy-back technique, like ISBX extension in MULTIBUS boards.

If each bus is confined to one connector only, and if the board carries two connectors, a possible mechanical structure is shown in Figure 6.2. With this solution all boards can be connected to either one or both buses. It is used for instance in M3 and in some MULTIBUS systems. In this last case the second connector, not specified by the standard, is user-defined. In M3 systems the two buses have the same specifications; it is so possible to fold the GB on one LB segment, as shown in Figure 6.3. Boards with only one connector, designed to work on LB, can also be tied to GB. This solution cannot be used for VME, because the second connector is already used for bus expansion and I/O.

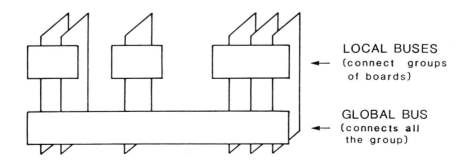

Fig. 6.2 – Mechanical structure for a hierarchical
        multiprocessor system.

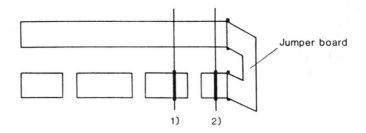

Fig. 6.3 - Folding of GB on LB

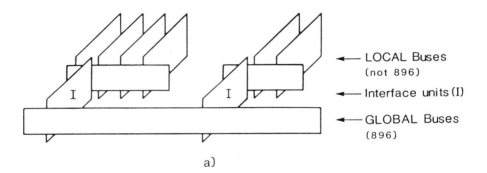

a)

- LOCAL Buses (not 896)
- Interface units (I)
- GLOBAL Buses (896)

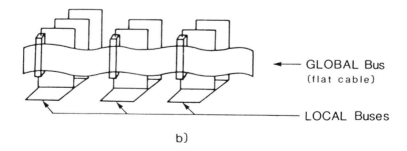

b)

- GLOBAL Bus (flat cable)
- LOCAL Buses

Fig. 6.4 - Structure for multiprocessor systems
        a) suggested for 896
        b) with flat cables

Another possibility, suggested for P896 systems, is to put the interfaces towards GB in a separate board, as shown in Figure 6.4a. Other simple solutions with flat cables have been used; an example |CIVE82| is given in Figure 6.4b. Flat cables are not as good as rigid backplanes for signal propagation: care must be taken in electrical interfacing (use of signal/ground pairs, shields).

6.2.2. Board Design Guidelines

The bus specification defines the electrical characteristics of boards and backplanes, such as voltage levels, input currents, and timings. When each module fits the specification, any subset of boards plugged into the same backplane is able to exchange information according to the bus protocol. As has already been outlined in Chapter 4, the propagation of signals on the bus lines, the loading of transceivers, and other effects can cause incorrect electrical behaviour (a "0" is sensed as a "1" and vice versa). These effects depend heavily on the technology used in the interface circuitry. An ideal interface should be able to drive a line independently of the load (that is, it should deliver any required current), and should read the state of a line without affecting it (that is it should have no input current). Existing transceivers however have limited output capability and non-zero input currents.

We must also consider that in many cases a bus signal must go to many different circuits on the same board. For instance, multiplexed DATA/ADDRESS lines in a slave unit should go to data buffers and to address latches and decoders. A line connected to many inputs sinks higher currents and, owing to the long tracks, can introduce reactive loading. To limit impedance mismatches and loading, each board should use a single transceiver for each bus line, located as near as possible to the connector, as shown in Figure 6.5. These bus transceivers make a buffer layer which uses devices optimized for bus driving and sensing, and isolates the board from the system.

In such modules the internal buses can be loaded (or driven) by many devices without affecting the external electrical behaviour of the board. It must be pointed out that the unity load interfaces may require in some cases more complex buffer control logic than multiple-unity load modules.

A disadvantage of the buffer layer is an increase of the signal propagation delay caused by the extra gates in the signal

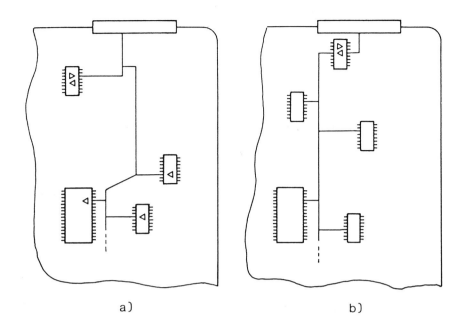

a)   b)

Fig. 6.5 – Board lay-out
              a) multiple transceiver
              b) single transceiver

path. This effect can be limited by tying critical signals directly
to the external bus, provided that each line is still loaded by a
single receiver. These solutions will be discussed in detail for each
module type.

The crosstalk between signal tracks can be limited by using
suitable layout techniques. The usual rules for grounding and
supply by-passing help to handle these problems inside the boards.
One must however consider that most of the crosstalk comes from the
backplane, where the bus lines run parallel and close to each
other for longer way. As was already pointed out in Chapter 4,
noise spikes are more dangerous for edge active lines. The lines
which must be protected, either passively (ground screens) or
actively (input filtering) are only those which carry edge-active
signals. Static lines, such as ADDRESS and DATA, which are strobed
by edges carried by other lines, must be settled and steady only in
proximity to the active edges of the command signals.

Good on-board supply decoupling helps to limit noise and
noise sensitivity. When provided (such as in M3, VME, and P896),

separate signal return tracks strongly reduce ground noise.

6.3. SLAVE MODULES

6.3.1. Organization of Slave Modules

A slave unit monitors the bus activity and,   when selected, becomes
a responder and can participate in a data transfer.   Typical slave
units are memory and I/O interfaces.   An intelligent controller or a
peripheral processor are also often considered by   the bus as
slaves.   We shall now analyze the organization of slave modules
pointing out the issues related to bus interfacing for basic
information transfer;   interrupt structures are examined jointly for
master and slave units in Section 6.5.
       As shown in Figure 6.6,   we shall consider the slave units
divided into two parts:

- CORE   : the device or circuitry which actually performs the
                task assigned to the unit.

- INTERFACE : the circuitry which links the core to the system bus.

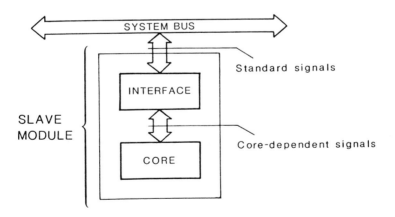

Fig. 6.6 – Basic slave unit

Typical cores are,   for instance,   memory arrays,   I/O registers,
peripheral   controllers.   The interface handles the bus handshake
and translates the bus signals into the signal required by the
specific core,   and vice versa.   Three categories of signals are

exchanged between the core and the interface: DATA, ADDRESS and COMMANDS, such as SELECT, READ/WRITE, STROBE, etc.

To examine in detail how each signal set must be handled, we shall further expand the interface submodule as shown in Figure 6.7.

A slave board can also contain many different cores. For instance, an ECC memory will have status registers mapped in the I/O address space, to report error syndromes. In this case the interface becomes more complex: it will contain two separate address decoders, two handshake circuits which may follow different timing, etc. The following Sections describe the organization of a simple interface, suitable for single logical slaves. The basic building blocks defined in this case are however still used for more complex circuits.

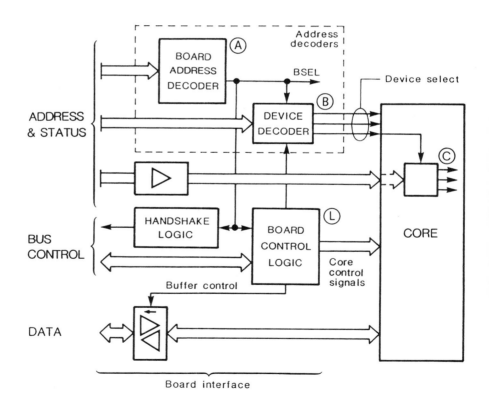

Fig. 6.7 – Block diagram of slave interface (non–multiplexed bus).

6.3.2. Address Decoders and Latches

In order to know if it has to participate in bus operations, a slave must recognize when the commander issues its own address by means of suitable ADDRESS DECODING logic. Usually the slave contains many levels of decoding, which corresponds to the ranks of board subunits. The board selection logic (A in Figure 6.7) recognizes the board address using status information and the most significant address bits. It activates lower decoding levels (B) and the board control logic (L), which in turn carries out the handshake, and the control of the core and of the buffers. The second decoder (B) selects among the on-board subunits (memory banks, I/O devices, etc.), using the adjacent lower address bits. The last level of decoding (C) occurs at the device level (e.g. inside a memory chip), and uses the remaining less significant bits of the address.

Address decoding logic can use comparators, gates, decoders, possibly buried into PROMs or PALs, as shown in Figure 6.8. It is good design practice to apply the ADDRESS VALIDATION signal at the lowest decoding level, to avoid the delay of ripple logic. Faster selection circuits can be obtained with parallel decoding, as shown in Figure 6.8e.

With multiplexed buses like P896 or M3, the address must be latched, either by means of transparent registers or edge triggered D-type flip-flops. Transparent latches allow anticipated selection of internal slave cells. In this case the address decoder must be disabled when the address can change, to avoid spurious selection spikes which are not acceptable in devices such as dynamic memories. Some examples of bad and good design are given in Figure 6.9. The latch can also be placed after the decoder, as show in Figure 6.9e. If N independent select commands are decoded from M address bits, and if $N < M$, this solution requires fewer flip-flops. We can examine as an example the organization of a 256 kbyte memory board for M3BUS, shown in Figure 6.10. It consists of 4 banks of 64 kword each. Address, data, and some commands are bussed to all memory devices, but each bank must have its own SELECT signal. If only one status condition and one address range must be selected, the board select circuitry can be accomplished using only a comparator or a decoder. If the board must be enabled on multiple status words or on many address ranges (e.g. a memory board with I/O registers), one must use a number of such decoders possibly buried in a single PROM or PLA. Some examples of multiple range selection circuits are given in Figure 6.11.

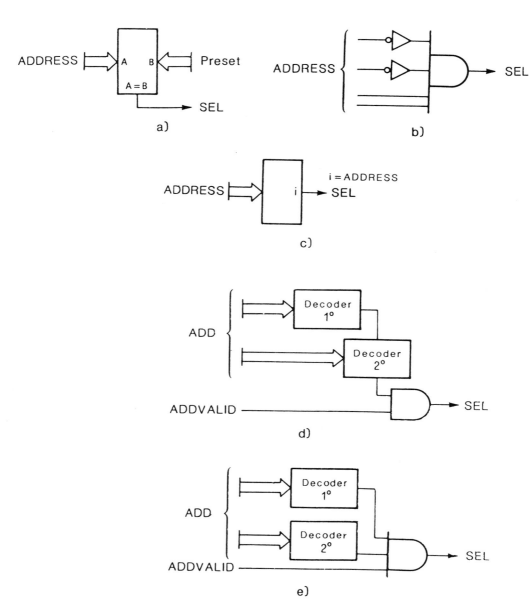

Fig. 6.8 – Address decoding logic
    a) comparator
    b) gates
    c) decoder
    d) ripple decoder
    e) look ahead decoder

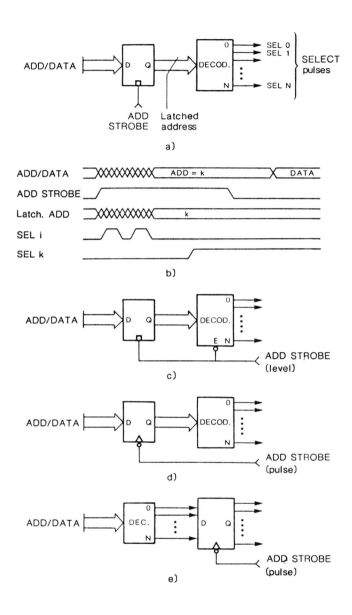

Fig. 6.9 – Address decoder and latches in multiplexed buses
        a) circuit which can issue spurious selection pulses
        b) timing with false selection pulses
        c), d), e) technique to avoid false selection

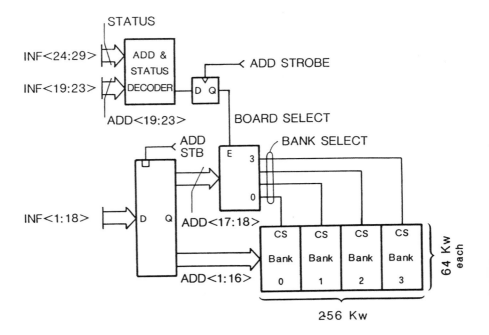

Fig. 6.10 – Block diagram of a memory module for M3BUS

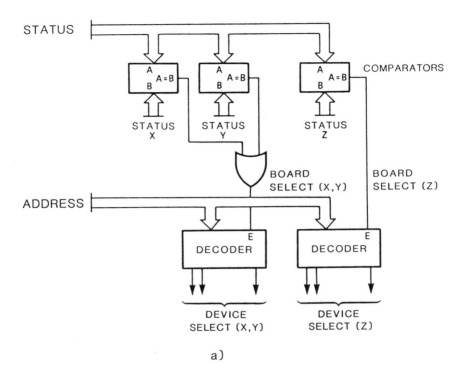

a)

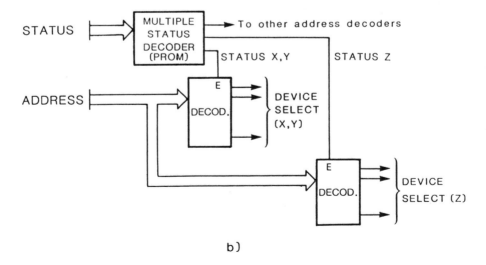

b)

Fig. 6.11 – Multiple status decoding
      a) Multiple decoders
      b) Single decoder

6.3.3. Slave Control Logic

The slave control logic, when enabled by the board address decoder, handles the handshake signals towards the external bus, and issues commands to the core and to the buffer/latch layer. The bus signals depend from the standard used for the backplane; all the others are more general.

An example of simple handshake circuit (suitable e.g. for MULTIBUS and VME) is shown in Figure 6.12. The delay td depends on the specifications of the core (access time for memories).

An example of more complex handshake logic which uses an E/D pair on the ACKNOWLEDGE action (M3 protocol), is shown in Figure 6.13. The delay t1 depends on the selection logic and must be as short as possible; t2 depends on the core of the responder; t3 and t4 depend on the commander. A suitable circuit to drive ACK-ENABLE and ACK-DISABLE lines is given in Figure 6.13b. Examples of other internal command signals for address and data cycles for the M3 protocol, are shown in Figures 6.14 and 6.15.

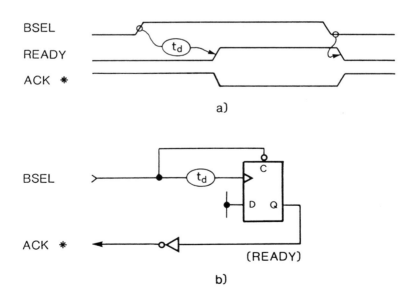

Fig. 6.12 - Handshake timing (a) and circuit (b)

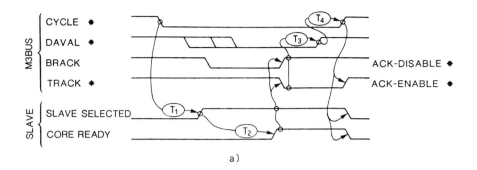

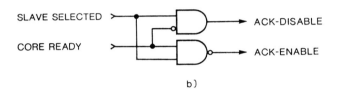

Fig. 6.13 – Handshake with the E/D pair
      a) Timing for M3BUS
      b) ACK–ENABLE/ACK–DISABLE circuit

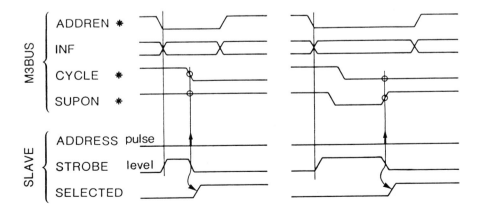

Fig. 6.14 – Timing of the Address cycle

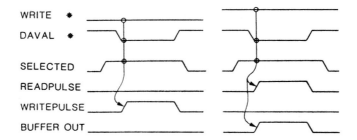

Fig. 6.15 – Timing of the Data cycle

6.3.4. Slave Buffering

The buffer layer isolates internal circuits from the external bus
and keeps the external "unity load" for any internal complexity. In
a slave unit we can identify three groups of buffers:

a) Bidirectional transceivers for data;

b) Line receivers for address and status, latches if multiplexed;

c) Line drivers and receivers for control signals.

In a multiplexed bus the part of address which is common to data
goes through bidirectional transceivers. In this case, to avoid
contention on the bus lines, the bidirectional transceivers must be
handled according to the following rules:

  - address cycle: enabled, input mode;

  - data cycle   : if the board is selected;

  - write operations: enabled, input mode;

  - read operations : enabled, output mode;

  - if the board is not selected: disabled, or enabled, input mode.

A simpler suitable control rule which works for both multiplexed
and non-multiplexed buses is:

- default: enabled, input mode;

- board selected for a read operation: enabled, output mode.

An example of control logic for data transceivers is   shown in Figure 6.16.

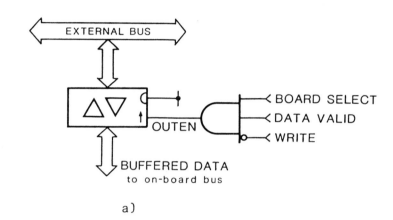

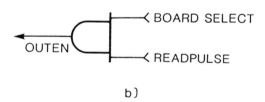

Fig. 6.16 – DATA Buffer control logic
a) Advanced write protocol (M3, VME, P.896)
b) Read pulse/write pulse protocol (MULTIBUS)

6.4. MASTER MODULES

6.4.1. Organization of Master Modules

A master module is able to begin and handle the bus activities, in order to transfer data to or from slave modules. Typical master units are processors and direct memory access (DMA) controllers.

It must be pointed out that, even if a master board   contains only
the CPU or a DMA controller, in most cases it carries also   some
slave units.   Here we shall only concentrate on the sub-units which
belong to the master logical structure, and which handle data
transfer and bus arbitration.   The slave sub-units are usually tied
directly to the processor, and are therefore independent from the
external bus.   In some special structures the internal slaves can
also be accessed from the external bus; they will be presented in
Section 6.6.   The interrupt control logic of both master and slave
units will be discussed together in Section 6.5.

     In order to outline the internal structure of master modules,
we shall examine the typical organization of a processor board,
shown in Figure 6.17.   This board can be tied to an external bus
and can handle information   transfers with other slave modules.
However, being complete of internal memory and I/O interfaces, it
can also work   as a stand alone computer.   This organization
facilitates debug and testing, because the processor board can work
either independently or plugged into a complete   system.

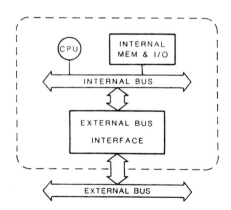

Fig. 6.17 - Organization of a processor board

In the structure of Figure 6.17, the CPU starts and handles all the
data transfer operations, and, following the definitions given in
Section 6.3, can be considered the CORE of the master.   When it
addresses the local memory or I/O, the external bus interface is not
active.   When the CPU selects a resource which does not belong to
local slaves, the external bus interface becomes active, and passes
the access request to the external bus.   In some cases, to aid the
debug phase, the interface can be forced to a "transparent" state,
thus making visible from the outside all the internal activities of

the board.

Internal access requests will be handled by local slave circuitry, which is usually designed to match the control signals of the specific processor. Since we are here interested in the structure of master modules, we shall concentrate on the circuit which handles external requests. As for slave modules, the signals exchanged between the core and the bus interface are: DATA, ADDRESS, COMMANDS.

The simplified block diagram of a master interface is shown in Figure 6.18. The CPU bus follows a manufacturer-dependent organization. It can be fully parallel (MC68000), or multiplexed (Z8000, 18086). In this last case, to interface standard memory and I/O devices, it must be converted into demultiplexed ADDRESS and DATA lines. We shall use this organization as a reference point for the bus interface. The internal bus shown in Figure 6.18 is, therefore, a full parallel bus with separate address and data lines, separate strobes and some kind of handshake.

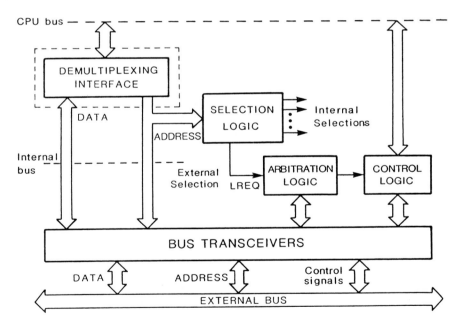

Fig. 6.18 – Block diagram of a master interface

If the board also contains other master or slave subunits, these will be connected to the internal bus Section, and from the bus the module looks like a unity-load master.

The selection logic discriminates between accesses to internal resources or to the external bus. Usually this circuit decodes the addresses of local memory or I/O and considers all accesses which do not select internal slaves as external requests. An external request enables the bus request and arbitration unit. When the bus request is honoured, the bus transceivers are enabled and a bus transfer cycle is initiated on the external bus.

6.4.2.   External Access Decoder and Bus Arbitration

The external access decoder is usually merged with the address decoder of internal slaves. Any access request which causes the addressing of a resource which is not on the board, is translated into an external acccess request, as shown in Figure 6.19. The selection logic can be organized following the examples given in Section 6.3.2.

    The arbitration unit can have different structures, according to the arbitration technique used. In a distributed arbitration technique, for instance, the same logic circuitry must be distributed among all the masters, while in a centralized arbiter the logic is assembled in a dedicated unit.

    Some examples of arbitration units are given in Figure 6.20.

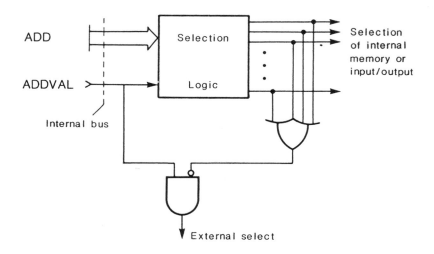

Fig. 6.19 – External address decoder

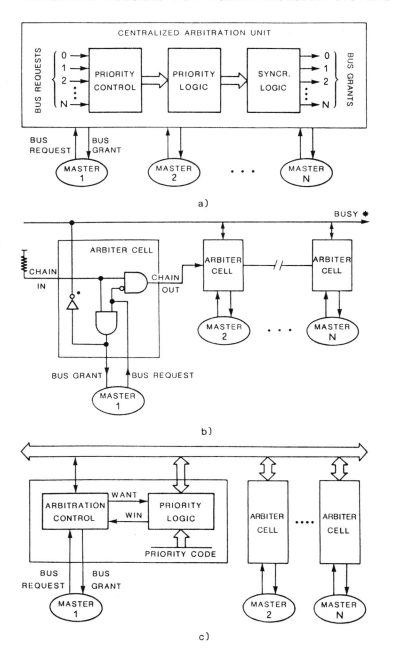

Fig. 6.20 – Block diagram of arbitration systems
   a) Centralized (MULTIBUS)
   b) Distributed, daisy chain (MULTIBUS, VME)
   c) Distributed, self-selection (M3,896)

6.4.3.   Master Control Logic

This unit handles the handshake with the external bus,   translates
the   control   signals   of   the   processor,   and   issues   the   proper
commands to the bus transceivers.
         As the CPU begins a transfer cycle,   the external selection
logic checks if the external bus must be accessed.   If so,   the local
bus request BREQ is raised and   the processor is blocked.   As soon
as the bus is granted to the master,   bus drivers are enabled,
address and commands go to the bus,   and the   cycle goes on
according to the sequence shown in Figure 6.21.   The majority of
integrated   processors   have   a   semisynchronous   handshake: the
transfer operation continues unless a WAITREQUEST becomes active
within a given time.   On the other hand backplane buses are
asyncronous.   The semisynchronous protocol can be converted into an
asynchronous one if the WAITREQUEST is activated at the beginning
of every cycle,   and deactivated by the READY response from the
selected slave.   A circuit which performs this operation is shown in
Figure 6.22.

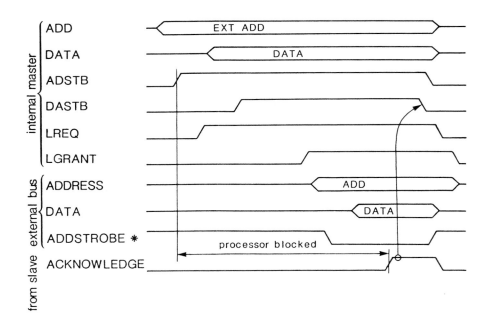

Fig. 6.21 – Interfacing of an asynchronous non multiplexed
processor with a multiplexed bus

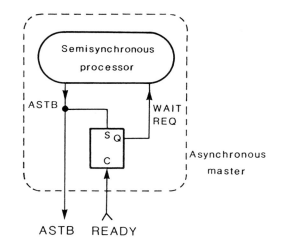

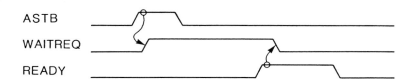

Fig. 6.22 – Interfacing of a semisynchronous processor
with an asynchronous bus

With the asynchronous handshake, when the processor tries to select
a slave which does not exist, no unit responds to the address in
the system, and the WAITREQ is kept active for an undefined period
of time.

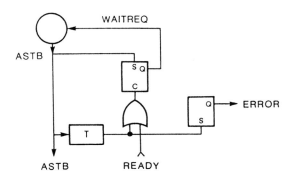

Fig. 6.23 – Block diagram of time-out logic

To avoid this lock situation, we must add a "timeout" circuit in order to allow the processor to complete the operation and to signal the addressing error. An example of timeout logic is given in Figure 6.23. When no slave activates the ACKNOWLEDGE handshake, a READY pulse is generated by the commander itself after a delay tw. This pulse resets WAITEREQ and sets an ERROR flip-flop.

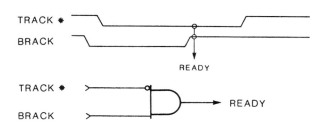

Fig. 6.24 - READY decoding circuitry (signal names for M3)

For instance, in M3BUS, the address phase proceeds synchronously, at a speed which depends entirely on the master. The data phase is asynchronous: the processor is blocked until the responder answers using the ACKNOWLEDGE ENABLE/DISABLE handshake. This condition corrisponds to READY active in Figure 6.22.

The double acknowledgement via the ACKNOWLEDGE ENABLE/DISABLE pair (used in M3 and P896) allows complete handshaked multiple-slave transfer; with the READY condition can be decoded as shown in Figure 6.24.

6.4.4.   Master Buffering

As was outlined in Section 6.4.1, a master board usually also contains slave submodules. In this structure we can therefore define three buses:

    - processor bus,
    - slave devices bus,
    - external bus.

Some possible interconnection schemes between these buses are shown in Figure 6.25 . More complex structures which also allow access to the local slave from the external bus will be discussed in

Section 6.6. If the slave devices are designed to match the processor structure, that is if they use the same protocol, the processor and the device buses can be directly connected, and the board contains only one interface from this processor/memory bus towards the external bus. When the processor and the slaves have a different organization, the board contains three distinct buses and at least two interfaces, as shown in Figure 6.25 b) and c).

The interface between a processor with multiplexed bus and the memory bus consists in most cases of a latch or register which demultiplexes addresses and data, as shown in Figure 6.26.

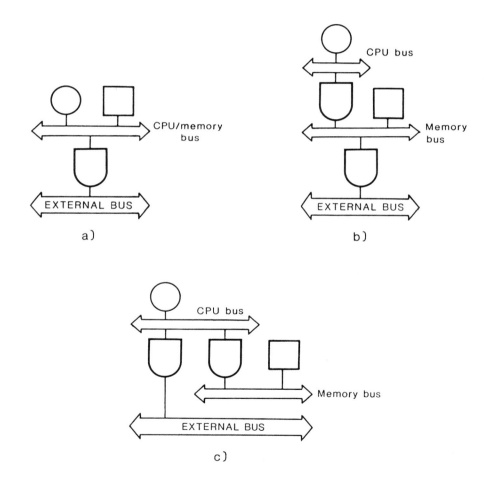

Fig. 6.25 – Bus structures for a Master module composed of a processor with local memory

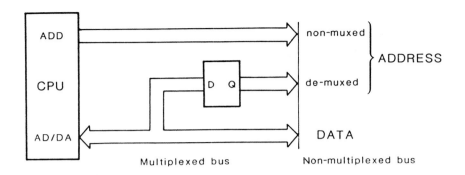

Fig. 6.26 – Processor–memory bus interface

Figure 6.27 shows two solutions to interface a parallel (non-multiplexed) bus with a multiplexed external bus. The parallel bus could be the memory bus or a non–multiplexed processor (e.g. MC68000) bus. A first possibility is to multiplex address and data using three state drivers as in Figure 6.27a. Since in this case two outputs go to the same Address/Data lines, to fulfill the electrical specification of unity–load boards, the multiplexing cannot occurr directly on bus lines. An additional buffer layer isolates the internal multiplexed bus and allows to connect a single transceiver to each line of the external bus.

The second technique, shown in Figure 6.27b, is to use a multiplexer and transceivers with unidirectional lines towards the board circuitry (half–bidirectional drivers). In this case the input data path is shorter and faster, because one buffer layer is saved.

If the CPU has an address/data multiplexed organization which copes directly with the multiplexed bus (e.g. Z8000 and other processors with M3BUS), the buffer structures shown in Figure 6.28a can be used.

Since in most cases multiplexed CPUs can insert wait states only in the data phase, if some cycles require the commander to stop in the address phase (all operation in P896, some supervisor operations in M3), we must again demultiplex and latch addresses, as shown in Figure 6.28b. If the processor can be stopped in the address phase, (e.g. by halting the clock), the interface towards the multiplexed bus is a simple transceiver layer, as in Figure 6.28a.

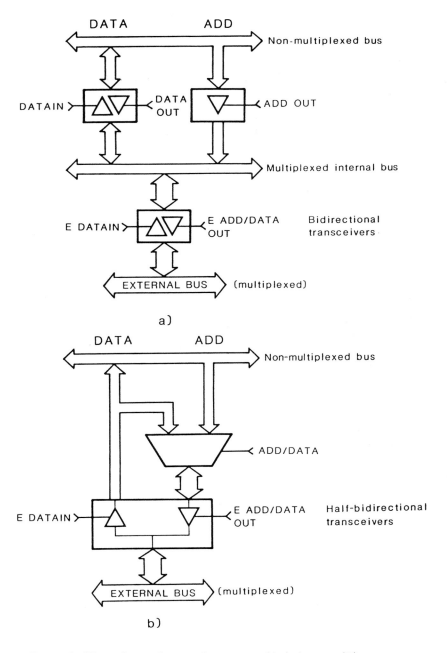

Fig. 6.27 – Interface of a parallel bus with
a multiplexed external bus
a) using 3-state drivers
b) using a multiplexer

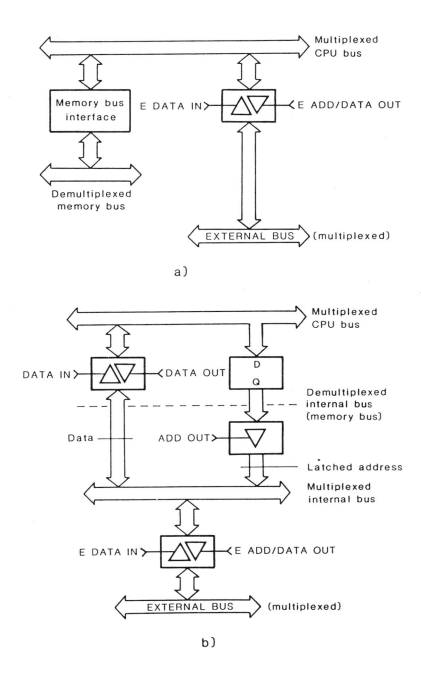

Fig. 6.28 – Interfaces of a multiplexed CPU with
a multiplexed bus

6.5. INTERRUPT STRUCTURES

6.5.1. Requirements for Multiprocessor Systems

An interrupt involves an information transfer which is not originated by the bus commander. We shall, therefore, consider as "interrupts" all information exchanges which modify the program execution and are not carried out following the usual sequence: bus arbitration, slave addressing, data transfer. They may be caused by attention requests from peripherals or other processors, bus errors, traps, memory violations, etc.
     In a multiprocessor environment we can identify various types of interrupts:

a) From a peripheral device to a well defined processor: this is the same kind of interrupt already used in single-processor systems.
b) From a peripheral device to all processors. It can be used to broadcast emergency signals, which request actions to be performed immediately (e.g. on power failure).
c) Specific processor-to-processor signalling, to support the exchange of messages between tasks. The term "specific" here means that the source already knows, and can specify, the destination of the message.
d) Generic processor-to-processor signalling. This involves information transfer between one source processor and one or more destination processor(s) which are not known by the source, and therefore cannot be "addressed". For instance, redispatch messages such as: "The processor running the lowest priority task must switch to task A" belong to this category. This last type of information transfer requires special hardware for self-selection of the destination |KIRR81|.

A multiprocessor system must support different interrupt techniques to allow the implementation of as many as possible of the features mentioned above. The minimum requirement is a "processor attention request" mechanism, which can be implemented at the hardware level by a simple interrupt-like structure (case a), and support other functions in the software. Usually some bus lines are reserved for system-wide emergency signalling, and this corresponds to case b).
     A hardware implementation of functions c) and d) allows more efficient multiprocessor/multitask environments; this direct hardware

support has been included in more recent devices such as I432 |I43283| and iAPX386.

In some cases (VME, P896, M3), backplane buses reserve lines for serial messages which can be used for this purpose. Namely, a serial line, besides providing an alternate path to the parallel bus, can be used to transfer special messages, with the possibility of self-selection of the destination, as required by example d). P896 uses only the serial line for all interrupt-like actions.

Each interrupt structure should be fully independent of the others. Since the external bus can be used at various levels in multiple-bus systems, the designer can use only the facilities required for the specific level, or share the same structure among different buses. For instance, in the TOMP machines |CICO83|, the device interrupt (case a) is used only on private buses, while the serial line connects all M3 buses (local and global levels).

In this Section we shall analyze the features and the implementation details of some interrupt mechanisms. The use of the different structures in multiple-processor machines will also be pointed out.

6.5.2. System Controls

Some backplane lines are always reserved for direct broadcasting of interrupt-like signals to all modules. A system control signal which broadcasts top-priority information to all modules and must exist in any system is the

RESET: when active, it forces all modules to a known state. On the active to non-active transition, it initiates startup procedures.

The information associated with other lines are for instance:

POWERDOWN: signals an event such as a power failure which must always change the execution flow in the processor.
                     This line is defined in MULTIBUS, VME, M3BUS.

PROCESSORDOWN:signals a local failure (e.g. a processor) in order to start recovery procedures (defined in M3BUS).

BUSERROR: signals that a bus error has been detected.
                   Parity errors are used in M3BUS and in P896 while
                   VME uses this line for a generic bus error.

The first two lines usually go directly to the interrupt inputs of
each processor. The BUSERROR signal goes to the processor through
an interrupt controller.

        Usually POWERDOWN is activated by a unique module, and no
hardware mechanism for the arbitration of multiple requests is
provided. PROCESSORDOWN is a wired-OR line,   activated by all
units;   the source is identified in the subsequent reconfiguration
process.

6.5.3. Processor Interrupts

A dedicated line (PROCINT) is defined in M3BUS for interprocessor
interrupt, so that a vector can be stored into the processor control
registers.

        The sequence of operations to issue a processor interrupt is
the following:

- The source processor sets up the vector in a set of output
  registers.
- When the vector is complete, a bus request is raised.
- When the bus is granted, a special cycle is started.
- During this cycle, the PROCINT line is pulsed to strobe the vector
  into destination registers.   The transfer is acknowledged via the
  normal handshake procedures.

The processor control register contains both direct commands and
source/destination/type identification fields.

        Figure 6.29 shows an example of processor control register
and a block diagram of the interface towards the CPU.   A flag
distinguishes broadcast commands, sensed by all processors, from
specific commands, directed to one processor only. To avoid loss of
data,   the processor control register must be read by the local
processor before the next processor interrupt operation on the global
bus.   It is the responsability of the user to guarantee correct
operation,   either by means of a queueing structure or by an
handshake mechanism.

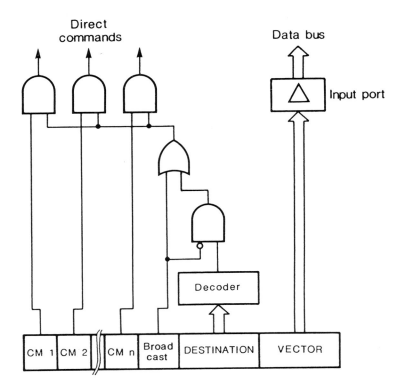

Fig. 6.29 – Processor Control Register

6.5.4. Centralized Interrupt Handler

Peripheral interrupts are mainly handled in a local environment;
they are addressed to a specific processor and handled via
centralized or distributed controller in the same way as for single
processor systems.

The problem is similar to bus arbitration; the controller must
discriminate between contemporary requests and allocate the CPU to
one of the requesters. This job can be accomplished using the
arbitration structures already shown in Figure 6.20. The CPU must
now also identify the requester to start the proper service routine.
Since the pure software approach, that is status polling, is too
slow for real-time environments, an identification vector is directly
provided, at the hardware level, by the interrupt control structure.

An example of centralized interrupt control structure is shown

in Figure 6.30.   This structure requires that each interrupt request
signal is activated by one device only,   and goes to a single
master.   The interrupt handler can be a simple priority encoder or
any   kind   of   centralized   interrupt   controller.   This   technique
associates a unique identifier with each request line;   the interrupt
circuitry converts the input requests into a vector which identifies
the highest priority active input.   This vector is read by the
processor and used to compute the address of the proper service
routine.   The   priority   of   each   request   can   be   modified   by
reprogramming the controller.
       A   centralized   interrupt   controller   is   used   for   non-vectored
interrupts in MULTIBUS.

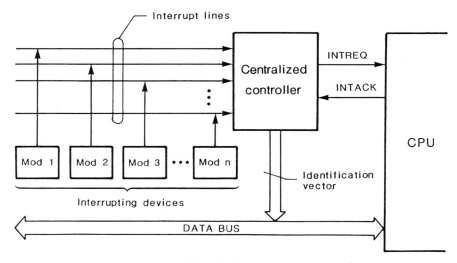

Fig. 6.30 - Centralized interrupt control

6.5.5. Distributed Interrupt Handler

In a distributed interrupt control structure, each requester provides
the identification vector. Arbitration is performed using daisy chain
or self-selection circuits, as shown in Figure 6.31.
       VME,   MULTIBUS and M3BUS use similar structures for vectored
peripheral interrupts,   the difference being the use of ADD/DATA
lines   and   the   daisy   chain   provided   in   VME   to   select   among
requesters tied to the same interrupt line.
       The peripheral interrupt structure of M3 is shown in Figure
6.32a.   Different subsets of INF lines are used to carry interrupt
requests,   identification of the acknowledged unit,   and the device
identification vector.

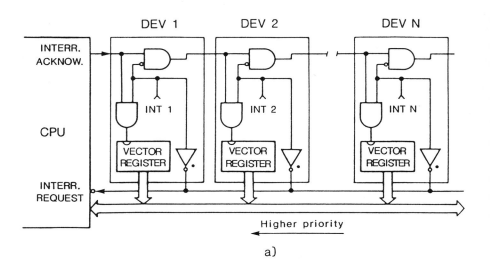

a)

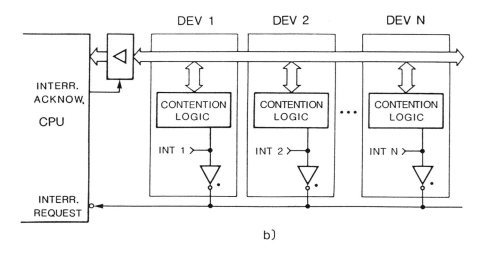

b)

Fig. 6.31 – Distributed interrupt control structure
          a) Daisy chain arbitration
          b) Self-selection arbitration

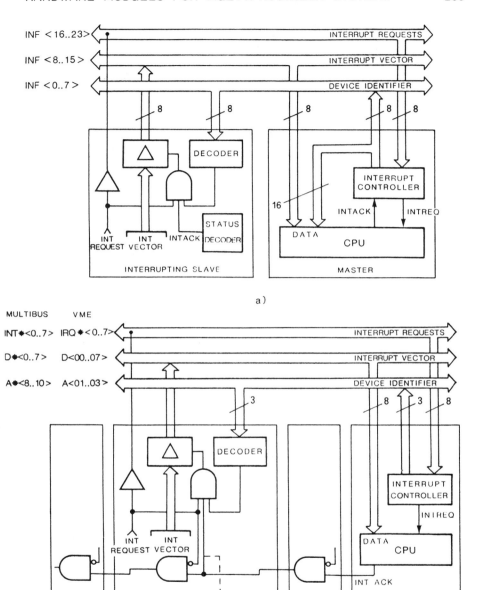

Fig. 6.32 – Organization of device interrupts
   a) M3BUS
   b) MULTIBUS, VME

In MULTIBUS and VME, the acknowledged device is identified using a subset of address lines. In VME many requests can be tied to the same line; they are arbitrated by chaining the acknowledge through requesters, as shown in Figure 6.32b.

With MULTIBUS and M3 many requests can be tied to the same line only if they come from a unique board, as shown in Figure 6.33.

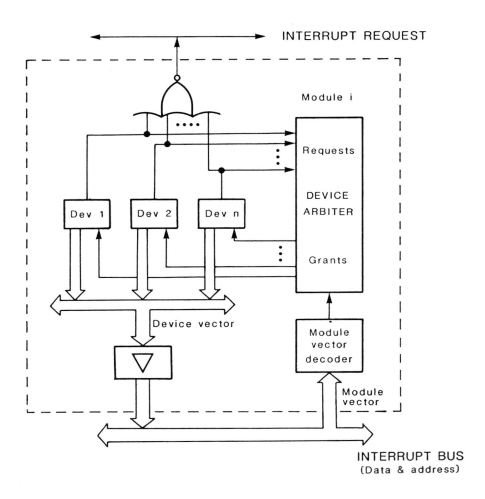

Fig. 6.33 – Arbitration of multiple interrupt
request inside a single module

6.5.6. Serial Lines

The serial line is the last technique we shall consider for exception signalling and handling. It provides a transfer mechanism which is fully independent from the parallel bus, thus giving the system designer the availability of two channels with different features. The parallel bus achieves high throughput by using a set of lines to carry many bits at the same time; its speed is limited only by the handshake. The serial bus, constrained to one line only, must multiplex the information and is, therefore, better exploited for the exchange of short messages.

The serial line controllers are handled by the processors as I/O devices. Some processing capability should be embedded in the controller itself, to identify the type of operations and issue the correct commands to the CPU. The organization of the serial line system is shown in Figure 6.34.

The use of the serial line is related to higher protocol layers which are not discussed here |KIRR81|.

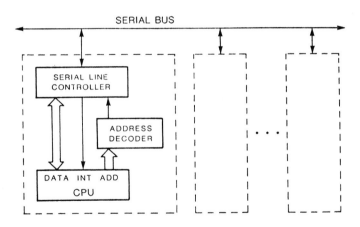

Fig. 6.34 - Block diagram of a serial line system

6.6. SPECIAL MODULES

6.6.1. Multiple-Slave Modules

Most slave boards contain many memory or I/O registers. This set of registers or memory cells is handled as a single logic unit by means of unique decoder and control circuits. True MULTIPLE-SLAVE modules are the boards which include many independent slave units, with independent addresses, and which in some cases perform different operations.

A typical example of multiple-slave boards is given by some complex memory modules. They contain programmable control registers to specify the operating mode (e.g. read only), and/or the bank address, and other on-board registers used by the error-recovery mechanism (syndrome register, fault address register). Such memory boards have two independent slave interfaces; one for the control and status registers, and the other for the memory banks. The module could be designed by simply putting the two slave circuitry on the same board; in this case there is no special problem, but some circuits are doubled and the module does not behave as "unity load" towards the external bus, because each line is connected to at least two transceivers.

A more correct approach is shown in the block diagram of Figure 6.35. A buffer layer interfaces the internal circuitry of the board to the external bus. In order to simplify the circuitry, some of the interfaces can be shared by all on-board slaves; for instance the same status and address decoders (usually implemented by means of PROMs and PALs), can select all slave units. The same applies to the handshake logic but we must consider that, in the general case, the delays associated with the two slave units are different. To handle the handshake with the same circuitry, the response delay should depend on the selected unit. The block diagram of a suitable circuit is shown in Figure 6.36.

Since the operations on the different slaves occur at different times, also the data bus and the address latches can be shared without interferences.

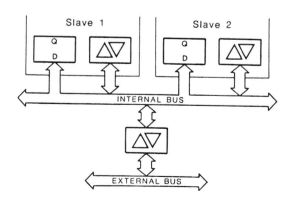

Fig. 6.35 —Multiple slave board

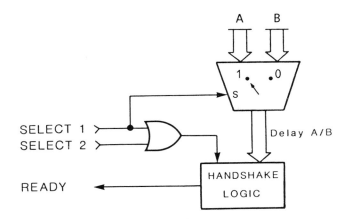

Fig. 6.36 - Handshake logic for a dual slave

6.6.2. Bus Windows

To extend the parallel bus beyond a single backplane, we can use
the two techniques shown in Figure 6.37.   The first structure uses
I/O interfaces connected by serial or parallel links.   The request to
access the register which holds the information to be transferred is
translated   by   the   commander   of   bus   A   into   messages   for   a
commander which is tied to the "extended" bus segment B.

        For instance,   to write a data in a memory cell,   the source
processor organizes   a message   which   contains   the operation   type,
address and data,   and sends   it   to   the   output   interface.   The
message is received and decoded by a processor or a DMA controller
on bus B,   which in   turn   executes   the   requested   operation.   The
sequence of operations in this case consists of the following steps:

a) information transfer into the output interface of bus A;

b) message transfer through the I/O link;

c) information transfer from the input interface on bus B.

Operations a) and c) use many bus cycles,   for message formatting,
I/O handling etc.

        The second structure,   shown in Figure 6.37b,   consists of a
direct link through a special interface unit,   which connects the two
buses by using only the buffers.

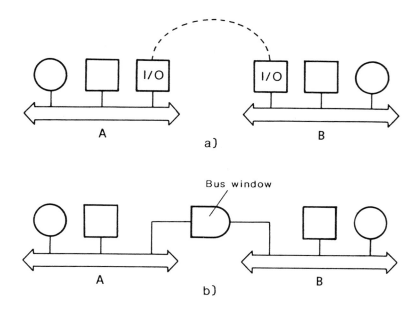

Fig. 6.37 – Bus extension
a) I/O loose connection
b) Tight connection through bus window

These interfaces are called BUS WINDOWs. The transfer in this case is performed in a single bus cycle and is fully transparent; the master on bus A handles it in exactly the same way as any other bus access. The sequence of operations for the access through a bus window is the following:

– The bus window recognizes that a slave tied to bus B is addressed;

– a bus request for bus B is raised;

– when bus B is granted to the window the transfer is carried on and acknowledged;

– the window is closed and bus B released.

As far as hardware is concerned, the first technique requires standard I/O interfaces only, while the second needs a special unit which is seen as a slave by bus A and acts as a master on bus B. Since I/O interfaces are simple slave modules, already examined in

Section 6.3, here we shall only analyze the structure of bus window modules.

The block diagram of a bus window unit is shown in Figure 6.38. When the commander on bus A makes an access request, the bus window must check if the addressed slave is located on bus A or on bus B. The address partioning between the two buses must be known to the interface. Any request for slaves tied to bus A is ignored by the bus window.

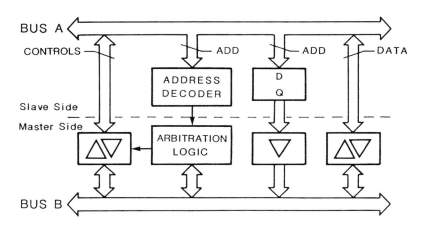

Fig. 6.38 - Block diagram of a bus window unit.
The latch is required for multiplexed buses

An access to bus B causes the activation of the master side of the module, which requests the bus through the usual arbitration procedure. The address on bus A is latched and the acknowledge is denied, thus forcing the processor on bus A to wait. When bus B is granted, the control unit of the interface rebuilds a complete transfer cycle. The acknowledge generated by the slave on bus B is passed to bus A. An example of complete sequence for M3BUS is shown in Figure 6.39.

A bus window interface must contain the complete circuitry of a slave towards bus A, and that of a master towards bus B. The timing signals for bus B must be completely rebuilt to guarantee hold and set-up times. It must be pointed out that the bus window is not symmetric, because it has well defined master and slave ports. When two buses are connected by a window interface, the information can be transferred in both directions using read or write operations, but the access requests can go in one direction only.

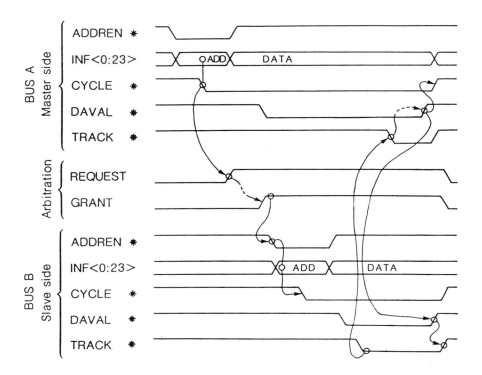

Fig. 6.39 - Timing diagram of a Bus window in M3 systems

A fully bidirectional bus window can be set up by cross-connecting two unidirectional windows, but is seldom used, because it goes to a deadlock state if both side try to go through the window at the same time. If the connection uses I/O links as in Figure 6.37, it can be symmetric or not depending on the software which handles the link. In some cases the bus windows also perform some address mapping, for instance by replacing the address field which identifies the destination bus with other information.

6.6.3. Dual-Port Slaves

The multiprocessor architectures described in Section 1 are built using master and slave modules which are connected to one bus only. Some of these structures however can be seen also from a slightly different point of view. Let us consider for instance the dual-port shared memory architecture shown in Figure 6.40. The dual-port structure is represented as a normal memory unit tied to

a bus which is connected by means of bus window interfaces to two
buses, but in most cases the memory and the bus interfaces are
built on the same board. This module corresponds to the structure
enclosed by the dashed line in Figure 6.40, and we shall consider
it as a single unit: a DUAL-PORT SLAVE. This grouping of basic
units can be extended to N-ports, but here we shall describe only
dual-port slaves.

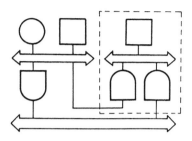

Fig. 6.40 - Dual port shared memory

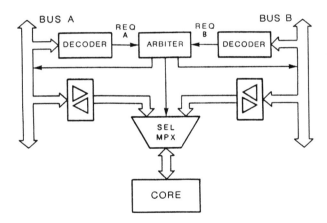

Fig. 6.41 - Block diagram of a Dual Port Slave

Small dual port memories are now becoming available as single
integrated devices. In this case the memory arrays have a special
organization to allow multiple read. The block diagram of a dual-
port slave is shown in Figure 6.41. It consists of two identical
slave interfaces, each complete with status/address decoder,
handshake logic, and buffers, plus an arbitration structure, a
multiplexer and a core. If the requests from the two sides occur at
different times, the multiplexer is switched in order to connect the

core to the requester immediately.   When the two access requests
occur at the same time,   the arbitration unit grants the core to one
port and forces the processor on the other side to wait.   The same
occurs if a request arrive while the core is busy.

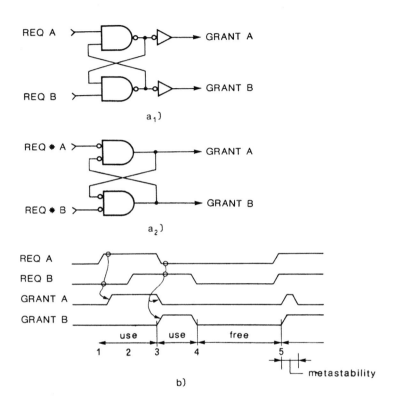

Fig. 6.42 - First-come-first-served two-input arbiter
                a) logic circuits
                b) timing diagram

The arbiter is based on a SR flip-flop,   as shown in Figure 6.42.
When   no   request   is   active   (input 0,0)   the   flip-flop   is   in   the
"forbidden" input condition and both outputs are high (no grant
active). If REQ A is raised   first, it sets the flip-flop in the state
which grants user A.   If now also REQ B is raised,   the flip-flop
goes in the "memory" input condition (1,1),   and the outputs do not
change.   When request A becomes not-active,   if REQ B is still
active, the flip-flop toggles and B is granted.

This simple two-input queue handler falls if the delay between the two requests lies in the same range as the propagation delay of the flip-flop.   In this case the feedback loop cannot settle before the next input change,   and the flip-flop may change state randomly for an undefined time,   or even go into a "fourth" state in which the two grant outputs may become temporarily active togheter,   as shown in Figure 6.42.

As an example to introduce this problem we shall consider a D flip-flop.   The behaviour of this flip-flop is specified when the inputs change one at a time,   and set-up and hold times are respected,   as in Figure 6.43a.   When two input (e.g. clock and D) switches close,   as in Figure 6.43b,   the flip-flop goes into an erratic undefined state,   called METASTABLE |CHAN72| |SAME82|. The flip-flop will finally switch to one correct state,   but the outputs can stay at undefined levels for an unknown time tm,   whose statistical parameters are related to the input timing and to the device technology.

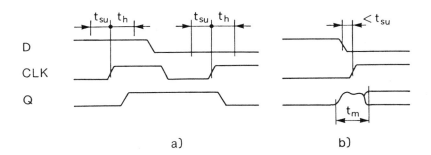

$$t_{su} \quad t_h \qquad\qquad t_{su} \quad t_h \qquad\qquad < t_{su}$$

D

CLK

Q

$$t_m$$

a)   b)

Fig. 6.43 – Metastable states in a D-FF.
        a) correct behaviour: Tsu and Th are respected
        b) insufficient Ts causes metastability of the FF

A possible solution is shown in Figure 6.44.   The delay reduces the probability of false grants at the expense of an increased arbitration time.

Integrated flip-flops are preferred for these circuits because, owing to the on-chip feedback path,   they have lower probability to stay into metastable state. However, in some cases the manufacturer does not guarantee the behaviour of the device in the "forbidden" input state,   and we must add input mask logic as shown in Figure 6.45.

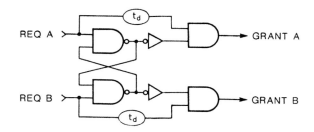

Fig. 6.44 – Filtering of metastable states
in the FCFS arbiter

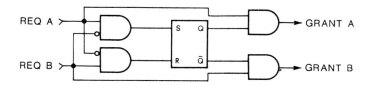

Fig. 6.45 – FCFS arbiter without metastable states

6.6.4. Master–Slave Modules

The master boards discussed in Section 6.4 contain both master and
slave units, but the slave can be accessed only by the on-board
master. In other words, the boards act towards the external world
only as masters. Here we shall discuss the organization of
modules which perform both as masters and as slaves towards the
external bus. A peripheral interface with DMA controller is an
example of true master/slave module. The controller is a master
but, at the same time, it contains control and status registers
which are loaded and read as slaves. Another example is given by
a CPU board with a local memory which can be accessed also from
the external bus.

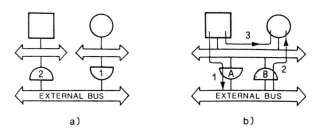

Fig. 6.46 – Organization of Master–Slave modules

Two possible bus structures for these boards are shown in Figure 6.46. It must be pointed out that in Figure 6.46a the module consists of two independent slave and master units with separate interfaces towards the external bus. The structure of each unit has already been discussed in Sections 6.3 and 6.4, respectively and the board cannot comply with the "unity load" electrical specifications. A merged master/slave structure is shown in Figure 6.46b. In this case a single bus interface connects the internal units to the external bus, and we can identify three different information transfer paths:

1) Master-bus , through the bus interface A;
2) Bus-slave, through the bus interface B;
3) Master-slave, without going out of the board.

The two bus interfaces are organized as bus windows; on the slave side an address decoder identifies if the access request must be passed to the other bus, while the master side must handle the bus arbitration mechanism. The block diagram of a complete interface for a Master-slave board is obtained simply by cross-connecting two bus windows as shown in Figure 6.47. Since this interface is actually a bidirectional bus window, care must be taken of deadlocks, as explained in Section 6.6.2.

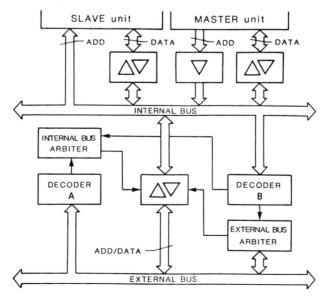

Fig. 6.47 - Bus interface for Master-Slave modules

6.6.5. Block Transfer Units

Some slave modules are able to transmit data at high speed,   using
the block transfer facility of some buses (VME,   M3BUS and P896).
During a block transfer cycle more than one read or write operation
is performed,   accessing consecutive memory locations in ascending
order.   Since only the address of the first location is sent,
multiplexed buses can almost double the throughput with this
technique.   Address is latched by internal circuitry and incremented
as appropriate.   Acknowledge signals must be sent for each data
item,   according to bus protocol.   On these modules also normal
transfers are allowed,   the transfer type being defined by the cycle
status signals.
      The general structure of the block transfer module is similar
to that of slave ones,   differing only in the interface submodule. In
this case the address latch   must be replaced by a counter as
shown in Figure 6.48;   this counter increments after each transfer
operation.
      In P896,   to speed up further the operations,   the block
transfer protocol uses the two edges of STROBE and ACKNOWLEDGE,
as in Figure 6.49.

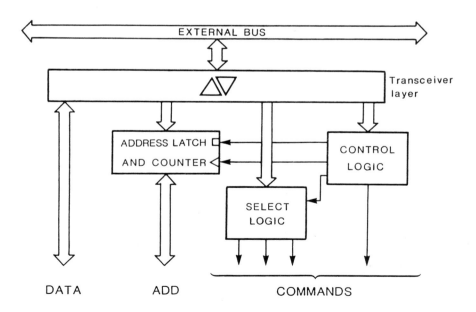

Fig. 6.48 – Block diagram of a block transfer unit

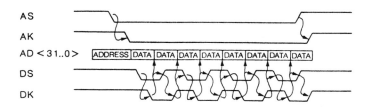

Fig. 6.49 – Block transfer protocol for 896

6.6.6. Supervisor Modules

As explained in Section 4.4.2, supervisors are special modules using the Enable/Disable technique for N-partner protocols. The operations performed are mainly of two types: monitoring and intervention. These operations cannot be accomplished by normal masters or slaves, but need dedicated units, able to slow, block or change a cycle by using the E/D technique on information STROBE (WRITEPULSE) and on OUTPUT ENABLE (OUTEN) signals.

According to the various operations performed, these units are organized in different ways; some blocks are common to the two basic structures which are described in the following. The examples shown in this Section concern mainly the M3 protocol, which has been designed to fully support bus supervisors. Some operations are possible also on MULTIBUS, which uses the E/D technique on memory selection, and on P896, which has a N-partner handshake and byte disable commands.

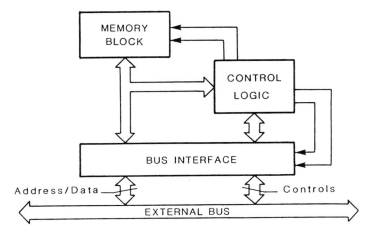

Fig. 6.50 – Basic operation unit

As shown in Figure 6.50, supervisor modules can be divided into three main parts:

BUS INTERFACE: it links the internal circuitry to the external bus.

CONTROL LOGIC: this circuitry controls all the internal operations.

MEMORY BLOCK : the part which contains the tables used by supervisor operation.

As for other units, a layer of buffers and latches is used to isolate the internal part of the module from the bus. This layer embodies both the master and the slave buffer layer. In fact, the module is not a simple master or slave, but it can act as a master or as a slave, according to the specific phase of bus cycle and the operation implemented.

This interface needs bidirectional transceivers on ADDRESS/DATA lines, since the module can replace, if necessary, the address as well the data.

The supervisor control logic handles the handshake with the bus, issues the proper commands to the buffer layer, to address latches and decoders and to the memory block. It contains the dedicated logic to support the operations specific of the supervisor.

In a monitoring supervisor, since no intervention has to be taken at cycle bus level, the logic has only to control the information flowing on the system bus. It can use interrupts for breakpointing, but the E/D technique is not required for these operations.

The Memory Block includes the tables and the I/O registers, which allow the processor to communicate with the supervisor. I/O registers are used for programming the module and for reading status information related to the functions implemented and with the operations performed.

An Address Replacement and Protection Unit (ARPU) block analyzes all the accesses to memory: unauthorized accesses are inhibited, by sending appropriate signals to the bus via the interface block.

The tables of ARPU units must use high speed memory, because the intervention is on-line and the bus operations must go on as fast as possible. In a monitoring unit, on the other hand, the memory has only to store the information flow and high speed requirement are not necessary.

An implementation example of supervisors can be found in the multiprocessor system TOMP |DELC83| |MINA83|, based on M3BUS. A subset of memory management functions is performed by more ARPU modules with address substitution and memory protection with write protection capability.

Violations are signalled to the processor using a non maskable interrupt. The module is then accessed as a peripheral device for reading the causes of the violation.

The complete block diagram of this supervisor is shown in Figure 6.51. The module contains the tables for both the violating handling and address substitution. The tables are divided into two different memory blocks, the last being used only during an address translation. Two I/O registers are used to store the internal tables by means of I/O operations. The first acts as a pointer to the internal memory and is automatically incremented to increase speed, while the latter really store the data.

The Protection Memory block is always used for violation detecting and contains the properties associated to each memory block. If address substitution is not allowed, the Relocation Memory and Relocation Control logic are not used.

The status decoder detects the type of the memory accesses and activates the Protection Control logic and, if enabled, the Relocation Control logic. If the access cycle can be completed, the system performs the address substitution as shown in Figure 6.52; otherwise the Protection Control block saves the causes of the violation, then ends the cycle and finally sends a NMI. During the interrupt acknowledge cycle, the Protection Control block sends an identifier that can be used by the interrupt routine.

The module interfaces with the bus with 7 I/O registers; four of them are directly used by the Operating System during the inizialization phase, while the other ones store the causes of the violation.

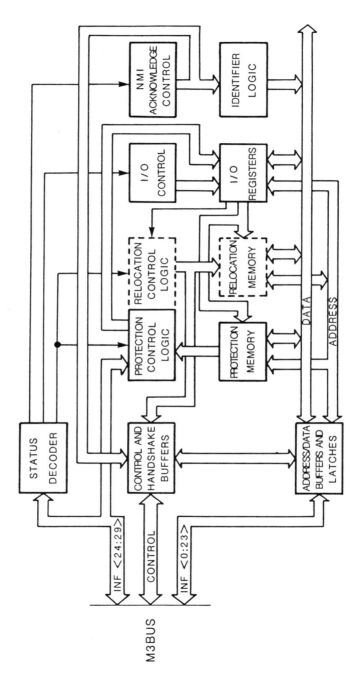

Fig. 6.51 – Block diagram of a TOMP supervisor unit

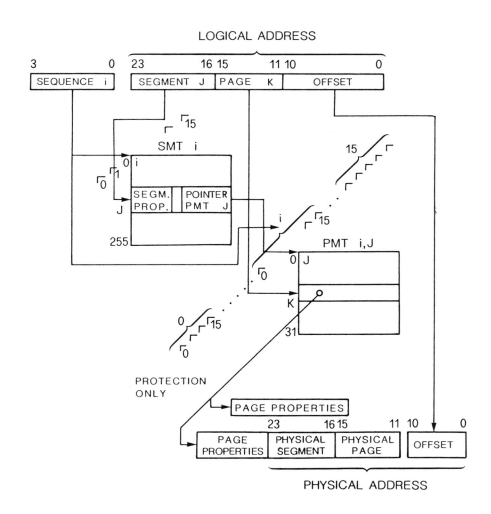

Fig. 6.52 – Address substitution mechanism in TOMP supervisors

6.7. REFERENCES

|CHAN72| Chaney T. et al, "Beware the syncronizer", COMPCON 72., 1972

|CICO83| Civera P. et al., "Multiprocessors: M3BUS systems and TOMP architectures" PFI-MUMICRO report, September 1983.

|CIVE82| Civera P. et al, "The u* project: an experience with a multiprocessor system", IEEE MICRO, May 82.

|DELC83| Del Corso D. et al., "Modulo supervisore per la gestione della memoria in sistemi M3BUS" AICA 83, Naples, September 1983.

|I43283| INTEL: "Microprocessor and peripheral handbook", 1983.

|KIRR81| Kirrmann H., "A serial interprocessor link for multiprocessor management in the P896 backplane bus", EUROMICRO 81 Proceedings, Paris, October 1981.

|MINA83| Minarelli G. and Zamboni M., "Estensione hardware dei protocolli di comunicazione nei sistemi a multiprocessore" Graduation thesis, Politecnico di Torino, October 1983.

|SAME82| Samet M. et al, "Metastable phenomenon in sequential circuits: simulation and prediction" AMSE Proceedings, Paris, July 1982.

CHAPTER 7

MULTIPROCESSOR BENCHMARKS

Eros Pasero
Dipartimento di Elettronica
Politecnico di Torino
Torino, ITALY

ABSTRACT.   This chapter presents some experiences of parallel
processing on a multiprocessor machine with shared memory (TOMP),
to extract some benchmarks. They are implemented using Sorting and
Merging algorithms and verify the operating mode of TOMP used as
a single board computer and multiprocessor system. A representation
form for parallel programs, named Parallel Notation Form, is also
presented. It has been used to describe the organization of the
algorithms implemented.

7.1. INTRODUCTION

A concept, generally accepted up to nowdays, is that the way to
increase the power of processors is to build faster devices. This
concept is now reaching technological limits; the speed obtained
from the fabrication process can not be increased by much more
without meeting physical limits.
     A possible solution to overcome this problem is to use groups
of processing modules to obtain machines whose speed and
reliability both result being superior to a high performance single
processor. In all the projects concerning very high performance
processing units, the actual approach to increase the speed is to
design the internal planning as a parallel architecture. However
the best possible exploitation of the parallel technique can be
obtained using an efficient parallel algorithm for the solution of
the given problem. By studying particular problems with algorithms
well fitted to multiprocessor machines, it is possible to obtain, even
with a rather simple multiprocessor system, a performance greater
than that of a sophisticated single processor machine.
     Our tests show that, when our algorithm is subdivided into

G. Conte and D. Del Corso (eds.), Multi-Microprocessor Systems for Real-Time Applications, 279–299.
© 1985 by D. Reidel Publishing Company.

many parallel tasks, the performance of the machine grows with the number of processors in a linear form, up to the limit of saturation of the system (see Section 7.2). This makes possible a comparison with a higher cost machine. Naturally, from these tests, the disadvantages of a multiple structure also come to light, giving a glimpse of its limits.

In the following Section the concept of performance analysis of a machine will be examined.

Section 7.3 deals with the concept of parallelism; it explains what is meant by parallelizing an algorithm, when it is possible, what are the advantages and what are the limits.

In Section 7.4 a new form of graphic representation of the parallel behaviour of the program is presented. It has been designed for the specific purpose of representing benchmark algorithms, but can be viewed as a general purpose tool for general use.

The algorithm we consider (Bubble Sort and Merge) is examined in Section 7.5. This algorithm is certainly neither new nor particularly efficient, but it gives the possibility to experiment various techniques for task synchronization to get the best parallel operation of the system.

In Section 7.6 the results obtained are presented in the form of tables and graphs. An analysis of these results is also presented, with a comparison against the results obtained on a VAX 780.

Finally, some proposals for future developments are presented.

7.2. THE CONCEPT OF PERFORMANCE

The first problem to be solved is to define what we mean by performance. Many definitions are in fact possible. |LORI72|.

One possibility can be the cost of the work performed by the system. However this parameter is more sensitive to the structure of the algorithm rather than to the real efficiency of the machine.

We chose to define the performance of a computer as the total amount of work that can be achieved over a given period of time. Therefore the parameter chosen to evaluate the performance will be the running time of the specific program. It should be added here that the performance evaluation using FLOP (Floating Point Operations per second) units, is not always satisfactory with parallel machines. In fact the gain of speed which results from increasing the degree of parallelization of an algorithm, can be greater than increasing the number of operations executed per

second.   For this reason we use a very simple algorithm which was
not very efficient in a sequential form,   but which exploits the
concurrency available in multiprocessor systems.

The performance of a computer depends both on the algorithm
and on the possibility,   offered by the architecture of the computing
system,   of an efficient implementation. The best fitting occurs when
the parallelism of the algorithm matches that of the machine.   In
this case the code produced will be a program of high performance.
The availability of a concurrent language allows more effective
implementations of parallel algorithms.   When such a tool is not
available it is necessary to use standard programming languages,
as Pascal,   for example,   and to carry out the mechanisms of
synchronization for parallel communication.   The main problem in
determining the performance of a multiprocessor system is finding
the factors that contribute to performance.   Contention on resources
(memory,   bus,   I/O devices,   disks) contributes to degrade the
efficiency of a multiprocessor.   The performance parameter would
have an optimal value if,   being T the time required to complete a
work on a single processor,   the time with a n-processor machine
were T/N.   Naturally,   owing to different levels of contention
introduced by multiprocessing,   this is not the case.   The limit of
saturation was studied by Brinch Hansen in |HANS73|;   it is
regulated by the formula:

$$N = 1 + Tl/Ta+Tg \qquad\qquad (7.1)$$

where N is the limit number of processors connected to the global
bus before it becomes saturated,   Tl is the time dedicated to the
local activity,   Ta is the arbitration time and Tg is the time
dedicated to the global activity.

It is therefore necessary to use suitable techniques to verify
these performances.   A good choice is the realization of a program
that can be executed,   in the same way,   by a growing number of
processors.   This approach gives real data on the speed-up obtained
by the increase of the parallel modules active in the system.   It is
also possible to compare the performance of a multiprocessor system
with that of a single-processor high performance computer. The same
programs implemented for TOMP,   used as a single board computer,
run on a VAX 780;   the results of this comparison are presented in
Section 7.6.

7.3. PARALLEL PROGRAMMING

Parallelism of an algorithm is defined as the number of independent arithmetic operations it contains and, therefore, can be executed concurrently |HOCK81|. Generally the best approaches used for parallel programming are the following:

a) the programmer can explicitly declare the parallelism using a concurrent language, as "Concurrent Pascal".

b) a compiler extracts the parallelism from sequential programs.

Not having any tool for concurrent programming, our choice was to implement a program written in sequential language, handling separately the parallelism . In a parallel algorithm the processes which constitute the algorithm must interact to effectively implement the parallelism concept to resolve the problem.

Processes can be divided into "Logic Steps"; these are the parts of a program that can be executed without interaction. At the end of every "Logic Step" the running process must communicate with another process; this is the "Communication Point".

The inter-process communication can be either synchronous or asynchronous |KUNG81|.

The synchronous case is when processes must wait for the completion of the logic steps of other processes before going from one logic step to another. Let us consider, for instance, the algorithm

$$A = B + C$$

where B and C are functions to be evaluated. If process 1 computes B while process 2 computes C in parallel, process A must wait for the completion of process 1 and 2. The disadvantage of this kind of processes is that, in most cases, it is impossible to forecast the time necessary for the completion of an algorithm. Every process has a data dependent execution time and all the connected processes are tied to these times. If a process enters a lock state, all the connected processes, waiting for its completion, are locked too.

In the asynchronous case processes have the following properties:

- they have access to a set of global , common variables.
- after every logic step, they modify the global variables,

on the basis of the finished work and proced to the following step.

This kind of connection is not as rigid as the previous case. The process has not as long to wait as in a synchronous algorithm; moreover there is no longer the risk that the lock of a process blocks the whole system |DIJK72|.

To evaluate the duration of the algorithm it is necessary to define a TOTAL TIME, which is the time elapsed from start to conclusion of the last finishing process. It is composed of three quantities:

- COMMUNICATION TIME, which comes from the inter-processes connections. In the synchronous form it is necessary either to set a semaphore or to send messages. In the asynchronous case it is necessary to set up a CRITICAL SECTION (which is explained in Section 7.4) |DIJK65|.

- WAITING TIME, with the synchronous communication, after every logic step, a process must wait for the synchronization with other processes. In the asynchronous case, a process may have to wait for the end of the activity of an other process on a shared resource (critical section).

- EXECUTION TIME, which is the raw running time of a program, without considering the interleaving of processes.

The total time is, however, a virtual time. It does not consider some characteristics of the system, that is the contention on the global resources and the possibility of saturation on the communication structures. In Section 7.6, the total time really measured with four processors and four parallel tasks is only slightly different from the theoretically foreseen total time. These criteria naturally change when the processor activities on the common resources increases, as seen in formula 7.1.

7.4. PARALLEL NOTATION FORM

A problem of graphic notation can arise in the description of parallel programs; the usual graphic form (flow chart) appears to be no longer adequate: the concepts of concurrency, synchronization and critical sections are not easily represented. For this reason we defined a set of formal and graphic rules that would allow us to describe clearly our parallel topics. Successively this analysis will be extended to study the realization of a tool of general utilization for the description of parallel programs.

The terminology used is defined in Figure 7.1, in a top-down schematization. We will call "PROGRAM" the parallel algorithm designed to obtain a specific goal. The program is divided into one or more "TASKS"; every task has a specific complex function. The task is divided into "PROCESSES"; they correspond, more or less, to Pascal procedures. A process is no longer subdividible (at logic level, of course) and it communicates with other processes, establishing a parallel environment.

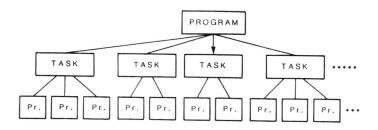

Fig. 7.1 - Parallel Program Organization

The parallel communication can occur in the body of a unique process or among different tasks. In the first case it is an inter-process connection; in the second case it is an inter-tasks communication. This second type of connection is organized with a process belonging to a process and the other process belonging to an other task communicating each other. There are two possible types of communications between processes: synchronous and asynchronous. In the first case, as already seen in Section 7.2, the process enters a waiting state, for synchronization with an other process. This form of synchronization can be a semaphore, a message passing primitive or an other form; these mechanisms are analyzed in |HANS76|. When process waits for a communication from two or more processes we speak of MULTIPLE SYNCHRONIZATION.

The asynchronous type is based on the concept of CRITICAL SECTION. It is possible that two or more processes try, at the same time, to modify the same shared data. If a process can access an area of memory where another process is working concurrently, some errors will result. Therefore it is necessary to control that processes will have mutually exclusive access to the set of shared data. These groups of data are the critical sections; they have these two characteristics:

    a) Only one process can be in a critical section at a time.
    b) Every critical section must be executed in a finite time
       and if a process attempts to gain access to a critical
       section, it has only a finite time for its attempt.

This last point avoids the deadlock that could occur with the
synchronous form. In fact, the main disadvantage of the
synchronous communication, is that, in the case of a failure, a bad
process would lock the whole system. In the asynchronous case, the
failured process could be switched off and the other processes can
continue their activities.

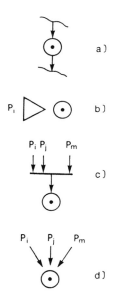

Fig. 7.2 - Synchronous Communication

Fig. 7.3 - Asynchronous Communication

The notations used for synchronous communications are shown in
Figure 7.2. The case a) represents the synchronous event: both with
a single and with a multiple synchronization the circle means that
the process is waiting for a synchronizing condition (WAIT state).

This condition can be represented by one of the three cases: b, c
and d.   Case b) shows that only one process must perform the
synchronization.   Case   c)   and   d)   represent   the   multiple
synchronization: in the first case all processes Ti, Tj ecc.   are
necessary to continue the activity (AND function);   in the second
case just one process is sufficient to exit from the WAIT state (OR
function).   The Figure 7.3 represents a critical section,   named Si.
At the two sides of the double arrow there are the names of the
processes which share the common variables.

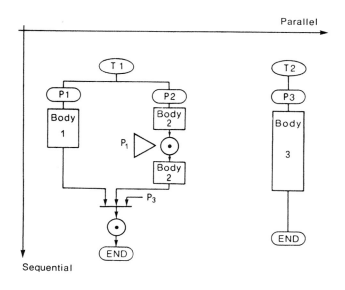

Fig. 7.4 – Parallel and Sequential Axes

If we look at the development of a program as depicted in Figure
7.4,   we can distinguish two axes which describe the type of
activity of tasks and processs.   The horizontal axis specifies the
parallelism;   the vertical axis specifies the sequentiality.   In this
example, tasks T1 and T2 are parallel: they are next to each other
on a horizontal line.   Inside task T1, processes P1 and P2 also run
side by side: they are parallel.   On a vertical axis: task T2 is
composed of process P3,   task T1 is composed of a set of two
processes,   P1 and P2.   Inside the environment of the processes, the
boxes represent the bodies of the processes: their organization is
strictly sequential.   Inside process P2,   there is an operation of
synchronization with process P1:   when the arrow enter the circle,
process P2 suspends its activity until it get the signal from P1;
then it continues.   The program finishes when the last active task

completes its activity, entering the END state. In Figure 7.5 a more complex parallel algorithm is shown.   Other examples are described in Section 7.5, with the explanation of the benchmark algorithms.

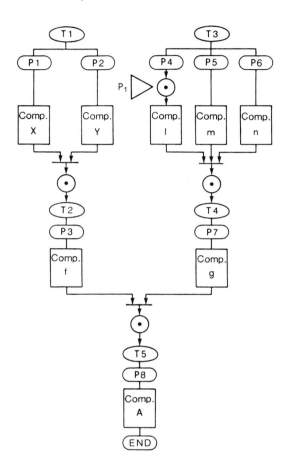

Fig. 7.5 – Example of Parallel Problem

Let us consider the function

$$A = f(x,y) + g(l,m,n)$$

where x, y, l, m and n are functions of other variables and l = h(x) is function of a function. The problem is subdivided into 2 parallel processes, P1 and P3; they calculate the internal functions of the two terms of the equation.   Inside the task T4 there is a synchronous communication: l = h(x) and x is calculated by P1. The task T1 finishes with an AND connection, which starts the task

T2. Likewise T4 calculates the 2nd term of the equation. Another AND connection introduces T5; this is the final task and gives the result. To add, for example, an other function inside the 2nd term of the equation, it is sufficient to insert a process inside task T3. To add a new term inside the equation itself, it is sufficient to insert a new task parallel to T4. There is a high degree of freedom: it is possible to change the algorithm, at task level, and to change the functions, (that is the operations performed by the algorithm), at process level. Parallel and sequential operations are easily identified by the horizontal and vertical directions. It is therefore easier to implement a scheduling philosophy for processes and processors. In the light of our experience this notation has been very useful. The adopted rules are however very essential; it should not be difficult to change them for more strict requirements of parallel programming.

7.5. PARALLEL SORTING TECHNIQUES

A usual technique for solving complex algorithms is to subdivide the original problem into a number of smaller subproblems and to obtain the final result from the solution of the small subproblems. |NIEL82|. Obviously the cost of the operation must be inferior to the cost of the original problem solution. This criterion is particularly attractive to implement a program for multiprocessor machines. In fact an increment of performance is obtained by increasing the number of processors, without any more construction. Now we examine some programs that implement Sorting and Merge algorithms. Sorting problems |KNUT73| have been intensively studied; a lot of effort has been dedicated to finding more efficient sequential algorithms. Recently the same problem has been studied for parallel machines trying to improve the algorithms. Here we study some available algorithms to explore the characteristics of multiprocessor machines. Table sorting is a typical problem that can be divided into many small parallel sortings, executed by the various processors. If contention and synchronization are not considered, the execution speed grows linearly with the increasing number of active processors, that is with the subdivision of the table to sort: the problem does not require any more effort. First a table of pseudo random numbers is generated. It contains from 500 to 8000 integer unsigned 16-bit numbers. The variable table lenght makes the experiment independent of the quantity of elements to be sorted. At this point the table is sorted with a technique of Bubble Sort and Merge. The reason for this choice is that, for what concerns performance, the problem is reducible to a raw bubble sort and the average time required is proportional to N^2, where N is the number of items to be sorted. If the table is subdivided into more parts, one for each processor, the performance grows linearly. Moreover the algorithm of Sorting and Merge runs, without any change, both with SISD and MIMD machines. So here we have a means of comparison to evalute the obtained performance. The random number table is divided into j parts, with j that varies from 1 to P, where P is the maximum number of processors available. The algorithms run on a multiprocessor machine, (TOMP), performing P parallel sortings and one final merging.

The sorting algorithm used is a "Bubble Sort". The reason for this choice is that here we are not interested to demonstrate the efficiency of the algorithm but we only want to show some benchmarks of general utilization, written using three languages: VAX-PASCAL, PLZ (a Pascal like) and Z8000 Assembler. Considering its single structure from a mathematical point of view, Bubble Sort presents higher visibility, which is very useful in the debug phase. It is important to remember that these benchmarks do not require any Operating System, or sophisticated Debug tools. Therefore an easily testable software is implemented. However, Bubble Sort is an autonomous module inside the program and it is possible to substitute it with an other more efficient algorithm such as a "Quick Sort".

Different types of benchmarks are implemented to exercise the system in various situations. PASCAL language is used for two reasons:

to speed-up the phase of writing the program;

to have a common tool for a large class of computers.

Z8000 systems, like TOMP, have available the PLZ-Z8000 language |PLZMAN|. It is totally compatible with PASCAL, for what concerns to our problems. Sorting programs for TOMP system are re-written using Z8000 Assembler |Z8000M| to optimize the code produced by the compiler and to increase the frequency of accesses to global memory, stressing the system. It is interesting, however, to verify the efficiency of the PLZ compiler: the redundant code is only twice the necessary minimal.

Another topic is to subdivide the table of the numbers to be sorted among 2, 3 and 4 processors (4 is the maximum number of processors actually available for TOMP). The first implementation

uses two processors;   the program structure is shown in Figure 7.6.
The table of random numbers is allocated in global memory,   in a
segment addressable from both processors.

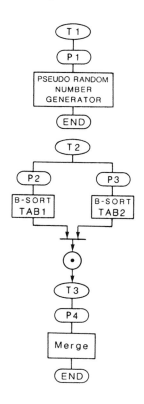

Fig. 7.6 – Two Parts Bubble Sort and Merge Algorithm

Two identical processes,   allocated on different processors,   sort the
two halves of the table.   At the end every process sets up a
semaphore;   this is the signal for an other task which merges the
two sorted tables.   This task presents, at the end, the initial table
sorted in a rising form.
The 3–processor implementation,   shown in Figure 7.7,   is straightly
more interesting.   The table,   divided into three equal parts,   is
sorted by three identical processs,   allocated on three different
processors.   Three semaphores are now used for synchronization.
When the sorting phase finishes a first Merge process (P5) merges
the first two semi-tables into a third table,   named Sj. At the same
time,   a second merge process (P6),   allocated on an other processor,
merges the third semi-table with Sj,   in a concurrent way with the

first Merge process. This form of communication is a type of critical section. The second merging process can take the items from Sj only after the other process has processed them. There is a test, before every access, if this condition is true. Moreover this process puts the merged table into the already existing three partial tables. Owing to the fact that the first merge process works on the same data, it is mandatory that P6 does not write on to the items before P5 can read them. This is a second critical section, named Si .

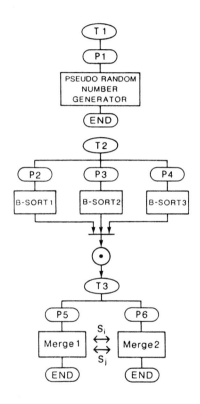

Fig. 7.7 – 3-Parts Bubble Sort and Merge Algorithm

The last implementation uses four processors,   as shown in Figure 7.8. Two identical processes, (P6 and P7),   allocated on different processors, merge the four sorted semi-tables into two tables. The last process (P4) merges the two tables into one. The subdivision into four tables is chosen to establish an omogenous algorithm for the performance evaluation. The programs ran firstly on a single processor; afterward they ran on two processors and finally on four processors. The same programs ran on a SISD machine (VAX 780).

It is interesting to notice that,   using the Parallel Notation

Form, the graphs already carry out the philosophy of the process scheduling. In Figure 7.6 the horizontal direction gives the degree of parallelism while the vertical one gives the sequential order.

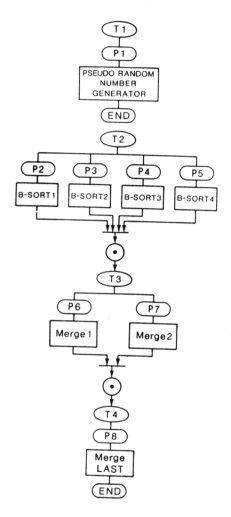

Fig. 7.8 – 4-Parts Bubble Sort and Merge Algorithm

Processes P2 and P3 can be implemented on different processors while P1 and P4, being rawly sequential processes, can only be implemented on the same processor. In Figure 7.7 the horizontal direction shows two degrees of parallelisms inside process P2 and process P3. Processes P2, P3 and P4 can be implemented on three processors and, likewise, P5 and P6 can be implemented on two processors. It should be noticed that, inside process P3, there are two critical Sections shared by the two Merge processes. A synchronous communication, (AND function), is also used: process P3 can not start until all three processes of process P2 have finished their activity. In the fourth algorithm, as shown in Figure 7.8, processes P2, P3, P4 and P5 of process P2 can be allocated on different processors. This is also the case for processes P6 and P7 of process P3. The other parts of the algorithm are sequential.

7.6. MEASUREMENTS AND ANALYSIS OF RESULTS

In these programs the code is strictly separated from data. Programs are resident in the private memory of each processor module, which is not accesible by other processors. The table of sorted items is instead allocated in the global memory, accessible to everyone. Four time meters are used to measure the total time. They are synchronized by the MO flag of the Z8000. Inside the program, a MSET instruction sets to "0" the pin MO of the Z8000 |Z8000M|. This signal is sent to the counter trigger input and starts the instrument. Likewise the MRES instruction stops the counting. The clock rate of the counters is 1 Mhz, allowing us to measure time in microseconds. Owing to the modularity of the programs it is possible to examine different solutions, varying the number of the subdivisions of the table and the number of the active processors.

The most remarkable result is the Assembler implementation with four processors and four tables. Examining the occupation level of the global bus we can notice an activity percentage of 70%, which is quite a high degree of occupation.

These benchmarks use two strategies to make the extracted times more reliable:

a) pseudo random numbers routine uses four different seeds for generations (3 for VAX);
b) five different quantities of items are treated.

From point a) it appears that the "disorder" generated by the routine is sufficiently uniform: the results are contained in a range of 4 %. From point b) it appears that the algorithm is independent from the number of items to be sorted. The number of items N and the number of parts P in which the table is subdivided are related with the sorting time T as follows

$$T = K N^{2/P}$$

(7.2)

K is a constant of proportionality. Table 7.I confirms that,
doubling the number of items, the times quadruple. Doubling the
parts in which the table is subdivided, instead halves the time.
This is the usual behaviour of Bubble Sort. The first information we
extracted is that the Merge phase does not influence the execution
time of the algorithm. All this demonstrates the evident superiority
of a Sorting and Merging algorithm in comparison with a simple
Sorting and, moreover, confirms the theory on parallelism given in
Section 7.3.

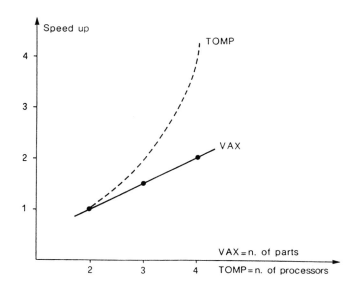

Fig. 7.9 – Speed-up for a multiprocessor VS.
a single professor with the Sort & Merge Algorithm

Figure 7.9 shows that, in a multiprocessor, (TOMP), the speed-up,
caused by the progressive subdivision of the table, is clearly
superior to that of an even more powerful single processor (VAX).
In fact a SISD machine is in accordance with law 7.2, with a
linear speed-up caused by the increasing of the subdivision of the
table. On the contrary, a parallel system uses a square law: there
is another speed-up factor nearly equal to the number of
processors. Naturally this demonstration is strictly tied to the type
of algorithm implemented. The problem must be carried back to the
parallelization of the algorithm. Only in the case of a perfect fit,
such as that here described, it is possible to show an advantageous
comparison between the two systems. If we define a
"disadvantageous factor" F as the rate between VAX and TOMP

performance,   using the same subdivision of tables (four semitables)
we have :

       F = 5.47 with TOMP used as SBC
       F = 1.47 with TOMP used as Multiprocessor (4 processors)

The resulting speed-up is 4.   One might think that, by doubling the
number of processors,   the increase of performance makes TOMP
clearly superior to VAX.   But Figure 7.10 shows that the speed-up
does not fit the theoretical curve exactly.

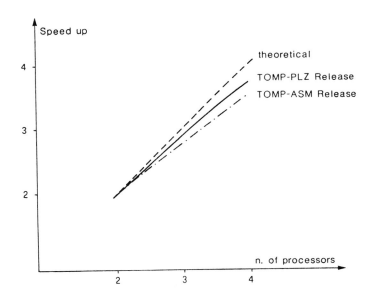

Fig. 7.10 - Speed-up of TOMP with a 4-Part Table
and a growing number of processors

In fact the performance of a multiprocessor are decreased by the
waiting time for shared resources;   with an intensive use the system
may also be saturated.   The theory described in Chapter 2, (formula
7.1),   already forecast this event.   Writing programs using the
Assembler language a more intensive utilization of common memory
can be obtained: the increasing of factor Tg decreases the value of
N.   Figure 7.10 effectively shows a soft decreasing of performance.
However we cannot really speak of decreasing in performance with

four processors: the curve is still quite closely fitted to the
theoretical one.

Table 7.I

TOMP. Bubble Sort & Merge of 1000 and 2000 items with a 2,3,4
parted Table and 1,2,3,4 processors. Z8000ASM Version.

Number of Items	Sorting & Merging 1 proc. 2 Tables	S & M 2 proc. 2 Tables	S & M 2 proc. 4 Tables	S & M 1 proc. 3 Tables	S & M 3 proc. 3 Tables	S & M 1 proc. 4 Tables	S & M 4 proc. 4 Tables
1000	15.567	8.026	4.137	10.040	3.653	7.566	2.183
2000	62.673	32.170	16.250	40.151	14.260	31.232	8.332

Table 7.II

TOMP. Bubble Sort & Merge of 1000 and 2000 items with a 2,3,4
parted Table and 1,2,3,4 processors. PLZ Version.

Number of Items	Sorting & Merging 1 proc. 2 Tables	S & M 2 proc. 2 Tables	S & M 2 proc. 4 Tables	S & M 1 proc. 3 Tables	S & M 3 proc. 3 Tables	S & M 1 proc. 4 Tables	S & M 4 proc. 4 Tables
1000	27.823	14.214	7.233	18.725	6.432	14.065	3.781
2000	114.175	58.602	29.410	74.548	25.758	56.973	14.824

Table 7.III

VAX.   Bubble Sort & Merge of 1000 and 2000 items with a 2,3,4
parted Table. Pascal Version.

Number of Items	Sorting & Merging 2 Tables	Sorting & Merging 3 Tables	Sorting & Merging 4 Tables
1000	5.140	3.4	2.57
2000	21.350	13.69	10.57

7.7. CONCLUSION

The aim of this experiment is:

- to make comparisons of performance for a multiprocessor machine by increasing the number of processors;

- to make comparisons of performance between a SIMD machine and a high performance SISD machine.

It must be pointed out that TOMP, in the Assembler
implementation with four tables and four processors, is faster than
VAX.   As a matter of fact we should say that the TOMP times are
heavily penalized.   In fact two WAIT states were added on the
global and private bus for technological reasons, so that the speed
of the system is halved.   Both program execution, that resides in
private memory, and global memory references are infact slackened.
Besides contention and waiting for resources, times shown in the
Tables 7.I and 7.II can be halved if no WAIT states are used.
Bubble Sort module can be replaced with other more efficient
modules (Quick Sort, Batcher's algorithm) without any modification
of the structure of the main program. Sorting time will therefore be
reduced , making the results more interesting from a mathematical
point of view. Another possible development can be the choice of
another problem, like the Fourier Fast Transformation or the
solution of differential equations, to verify the performance using
other algorithms. All these problems can be infact parallelizable
and can allow us to check different operating conditions. The most

attractive test, from an architectural point of view, is the increase in the number of processors. In chapter two the study of a curve similar to the one shown in Figure 7.10 gives a levelling of the curve with six active processors. It means that not all processes are fully active. There is even a study |DEMI82| that shows an inversion in the curve: the performance of the whole system decreases using more than six processors. It could be interesting to find the maximum performance of the tested multiprocessor machine, for a range of applications, versus the number of processors.

7.8. REFERENCES

|DEMI82| Deminet,J   "Experience   with   multiprocessor
              algorithms", IEEE Trans. on Computers, vol.C-31 n.4,
              April 82.

|DIJK65| Dijkstra, E. W., "Solution of a problem in Concurrent
              Programming Control", CACM, vol.8 n.9, September 1965.

|DIJK72| Dijkstra, E. W., "Hierarchical ordering of sequential
              processes", Operating Systems techniques, Academic
              Press, London, 1972.

|HANS73| Hansen, P. B., "Operating Systems principles", Prentice
              Hall Series in Automata Computers, 1973.

|HOCK81| Hockney and Jesshope, "Parallel computers", A. Hilger
              LTD, 1981.

|KNUT73| Knuth, D. E., "The art of Computer Programming", vol.
              3, Addison Wesley, 1973.

|KUNG81| Kung, H. T., "Synchronized and asynchronous parallel
              algorythms for Multiprocessors", Tutorial on parallel
              processing, IEEE Computer press cat. n.   EH0182-6,
              Bellaire, Michigan, 1981.

|LORI72| Lorin, H.,   "Parallelism in hardware and software",
              Prentice-Hall series in Automatic computation, 1972.

|NIEL82| Nielsen, and Staunstrup, "Early experience from a
              multiprocessor project", Aarhus University, Denmark,
              1982.

|PASE81| Pasero,   E.,   "A Multiprocessor Monitor Debugger", acts
           of IEEE conference, Toronto, 1981.

|PLZMAN| ZILOG, "PLZ user's guide", 1980.

|Z8000M| ZILOG, "Z8000 CPU Technical manual", 1979.